The Myth of Security

Hackers' Inventions Will Win The Race for Information

Niels Starling

To technology and its culture.

Table of Contents

Introduction

Do you feel stressed when you go online, thinking your personal information is at risk? Does the internet seem like a minefield, where you must be super-careful to avoid all threats? Of course, it is always better to be safe than sorry. But is the web really so dangerous? Sometimes we get bombarded by unsettling news, where every single incident is displayed way out of proportion. As a result, most people form a distorted perspective of life online, becoming evermore terrified to collaborate and cooperate with others over the internet. Does that sound familiar to you?

Time and over again, the media tends to focus on data breaches that take place online. Obviously, there have been several breaches of personal information over the years. In 2019, a massive data leak was brought to light regarding Facebook and 419 million users, along with their unencrypted Instagram passwords (Winder, 2019). One year earlier, it was Marriott that had to admit to a massive data breach from its pristine Starwood guest reservation software (Gressin, 2018). Back in 2019, Microsoft left its customer support database without proper security for almost a month and the repercussions were huge (Cimpanu, 2020). And this is just the tip of the iceberg. There are hundreds of other cases, where hackers breached the security measures of even the most prestigious companies worldwide. They reached their servers and got their hands on some precious data.

Often these breaches focus on companies that have vast amounts of consumer data. Those companies are willing to spend large amounts of financial resources to keep them secure and, in many cases, pay a ransom to get that information back if stolen. Hackers target big tech because it means big payouts. This doesn't mean each individual user is seen as an attractive target for an advanced hacker. Still, people create this distorted view that "if big companies can't even keep their data secure, how am I supposed to?" So an update on Apple or Facebook being hacked serves as proof of how vulnerable security is online. Such news is indeed threatening, but it doesn't mean these leaks are common. On the contrary, they are quite rare and only affect a fragment of the information that is shared worldwide. With the outstanding progress of technology, the internet is now better shielded than ever before against malicious intent.

It goes without even saying that all internet users must be educated on the kind of information they should never share online. Sharing their home address or the location of where your children attend school is not the right way to go. The same goes for your credit card credentials or your mobile phone number. You would not share that with a total stranger in real life, would you? Instead, you should filter the information and only publish what you wouldn't have a problem letting the world know. Even then, make sure that you practice safe web browsing and the risk is instantly lowered.

That being said, there is no point in overstressing about your online presence. When using an avatar to connect to a forum, the chances of exposure are near nonexistent. In a similar pattern, you can choose to use an alias as a username rather than your own name. This will allow you to express yourself more freely and minimize the risk of identity theft. Plus, you ought to be careful when adding friends and followers on social media platforms. If they look fraudulent or untrustworthy and ask you to connect, don't. Furthermore, adjust your privacy settings in a way that enables you to stay away from prying eyes and unwanted requests.

One of the pillars of online security is undoubtedly the use of unique, hard-to-guess passwords. So if you are concerned as to the vulnerability and frailty of your online presence, then be sure to step up your game. Select passwords that do not define who you are or what you like. In fact, there are cool apps that help you to generate and protect all your passwords. They also allow you to store them safely, keeping your mind at ease. Among the most popular apps, you can try are 1Password (*1Password*, 2013) and LastPass (*#1 Password Manager & Vault App, Enterprise SSO & MFA | LastPass*, 2019). Find out which one works for you.

Is Open Information Vulnerable?

People hesitate to embrace P2P sharing and open source information because the news outlets focus on security breaches and "information vulnerability." This creates widespread fear that makes internet users avoid truly creating and sharing online. However, things could not be further from the truth. P2P is just a vessel, a means to help internet users make the most of their online experience. Through the process of upping your online protection, you automatically enhance your shield against any malicious attempt. This is where you need to focus, rather than identifying the web as a scapegoat.

As soon as you realize what you must protect online, there is nothing stopping you from experiencing the web in the way you are entitled to. Why shy away from the challenges that arise online? More than that, why not benefit from the amazing wealth of information and sharing? Pure magic happens online, when you free your mind from prejudice and fear. Instead of avoiding surfing the web because of the dangers that lurk in the dark, shed light into these threats and address them for what they really are. Knowledge assists you to see more clearly and enjoy the internet without any hesitation.

Throughout the rest of this book, you will get to see fascinating examples that unveil the beauty of open source and effective information sharing. You will be able to see how people from different backgrounds have managed to work together with a common goal in mind. Indeed, such a wonderful aspect of the internet should be part of the human experience. It is the quintessence of solidarity and unconditional help from one person to another. Without any hidden agenda, without any obstacle getting in the way, P2P offers people the opportunity to share.

Other than in a serious threat within the real world, withholding information is a poor strategy in negotiation; if anything, the main strategy is to buy yourself more time, not provide the criminal with little information. In proper negotiation, both parties want to inform each other of what they expect on their end of the deal. Through this process, new information arises that was unknown to both members and this creates amazing collaboration opportunities, or "Black Swans." This can be seen in Chris Voss' amazing book *Never Split the Difference*, where he talks about the theory of Black Swans. "Black Swans are just a metaphor, of course. Think of Pearl Harbor, the rise of the internet, 9/11, and the 2008 banking crisis. None of the events above was predicted—yet on reflection, the markers were all there. It's just that people weren't paying attention. ... The Black Swan symbolizes the uselessness of predictions based on previous experience. Black Swans are events or pieces of knowledge that sit outside our regular expectations and therefore cannot be predicted." (Voss & Raz, 2016)

Look at the P2P revolution and take Napster as an example. This is nothing short of a Black Swan. The world needed such an innovative tool in their hands. They needed to be able to share music for free, without limitations as to the volume of the shared content or the financial status of those enjoying the music. Still, besides the true connoisseurs in technology, nobody could have seen that coming. Sharing online has become easy and efficient, thanks to the vision of the few that paid attention to what everyone wanted and needed. With the increasingly fast speed of the internet, it now sounds self-explanatory why P2P is more useful than ever.

Alas, there are forces that prevent online sharing from rising to its glory. When people worry too much about a potential data breach, eventually they lose focus. They sacrifice the awesome benefits deriving from P2P and avoid it, simply because of these unnecessary fears and over-exaggerated threats. Instead of taking full advantage of the web and its functions, people often sabotage their progress and get left behind, ironically thinking that they are doing what is best for them.

Relax and Enjoy the Ride

We have already referred to the safety practices that every user needs to comply with online. It makes sense why you should not reveal any personal information on the internet. Furthermore, it is essential that you safeguard your data through the use of password management apps. These are all elements that add to your overall sense of security. The internet is not the enemy, nor should it be considered as such. Once you get the hang of such practices, you will notice that a huge burden has been lifted off your shoulders. Sigh with relief and enjoy that feeling of complete online freedom. Take a nap in the P2P hammock—it's all good from here on out.

The comfort of internet security will take so much stress off your shoulders, you'll begin to wonder why you had so much stress before when you always felt your information was at risk. Imagine that your internet browsing experience was just a pile of stress and potential threats piled atop other stresses and threats. That would be an awful way to spend your precious time. Shouldn't "social" media be something to enjoy? Well, Facebook has implemented algorithms that identify your emotions and exploit them for their own gain. Does that sound absurd? Take a moment to think about the ads that appear on your newsfeed. Do they have anything in common? If you look closely, you will realize that the articles and headlines, the comments, the videos and the sponsored ads have been personalized per your own preferences. In case you are wondering how this can be, the algorithms are quite thorough. They gather every single like and every comment, measuring your reactions and quantifying them (Rose-Stockwell, 2017).

Do not let that scare you! As mentioned earlier, you can modify your privacy settings online and you can exclude ads from appearing on your newsfeed. Even in this case, avoiding social media platforms like the plague is not the answer. Sure, there is an entire industry that has been built around the interpretation of your feelings and attitude towards things. But this is roughly called marketing and should not be an obstacle in your attempt to experience online life to the fullest. Social media comes with a great many advantages which should not be overlooked. What you ought to do, however, is learn how to balance online safety with a positive and pleasant outlook toward the internet.

By reading this book, you can start to change your mindset about what you perceive as online threats. You will find it interesting to evaluate each so-called threat and see if it is indeed something to stress over. At times, you will notice the things that frightened you the most were the ones that now made you laugh. Once you get things straight in your mind, you will see that a whole new world unfolds before your eyes. A digital world that is filled with surprises and possibilities. You will become a more relaxed internet citizen, by embodying these philosophies into your daily life. No more fear of the unknown that lies ahead!

The quicker you adopt a positive philosophy about the web, the more benefits you will get out of your internet experience and your real life. You will gain a better understanding of the exciting opportunities that P2P can offer and where to find them happening today. Moreover, you will realize that some concepts cultivated from internet enthusiasts are actually the pillars of an ideal modern society. Slowly but steadily, you will eventually create a balance of minimal internet stress and maximum life comfort. Isn't that a match made in heaven?

The P2P Revolution

It is mind-blowing to think of free online file sharing. We have P2P technology to do that thanks to the early efforts of Napster in the late 90's. Back in 1999, two teenagers envisioned a new way of using the web and they made it happen, no matter the obstacles that they had to face. Back then, obtaining music meant that you had to buy an entire album on CD, vinyl or cassette. This made it a lot more difficult to have a wider view of music, limiting your options solely to the titles you could afford. Who would buy an album just based on a hunch? Thanks to that revolutionary concept, music became more accessible and artists could promote their work to a far larger audience than before. Open source file sharing of .mp3's became the new norm and people flooded the internet to gain access to these vast musical libraries (Barnes, 2015).

P2P creates a network of computers and these computers interact with each other. Do you fully comprehend what this means? This concept is so exciting that it makes my heart tickle. My computer may be located in the U.S., packed with a rich collection of country music in MP3 format. The files are classified by music genre, artist, and date of release. They also contain some details regarding the specific quality of the audio or the exact concert they were recorded at, etc. Having done all that, I have neatly stored them and they are readily available for anyone to check out.

Now a girl from Sweden goes online and wants to listen to some music. She wants to experiment with something that is outside her comfort zone. So she picks folk music and ends up searching for country songs. Finally, she bumps into my collection and gets intrigued. She wants to download these songs and give them a try. She clicks a button and there is it! After a while, she has a nice collection of songs that can keep her company, courtesy of a fellow country music enthusiast from the States. Perhaps I have subconsciously influenced the music taste of an individual with whom I have never interacted before in my life.

In a nutshell, this is the way P2P works, under the radar, without fancy advertising or false promises. Obviously, this is not the same as buying the real stuff. Nobody can claim that a censored studio recording will sound as nice as the end product. But in terms of rarity and exclusivity, sometimes the former outranks the latter. Furthermore, there are many other things to share online besides music. You can share that great TV series you recorded on your VCR back when you were just a kid. Maybe you want to share some manuscripts that have come to your possession by chance. Or, perhaps, you want the world to read all about the work of Voltaire or Aristotle.

True giants in the niche of innovative technology like Tesla are advocates of the same vision for collective information sharing for free. Software updates are always offered to Tesla vehicles, enhancing their performance and improving their overall security. All that happens via Wi-Fi, using the internet as a means to optimize the experience buyers get from their products (Software Updates, 2019). At the same time, Elon Musk promotes sustainable transportation, as well as eco-friendly innovations for households and businesses alike. As you can see, P2P sharing is a tool that the most brilliant companies use to their advantage; all the more reason to celebrate and embrace.

I want to take a brief moment to admit that these methods and technologies can be used for malevolent purposes. There is no argument in that, since terrorists have been taking advantage of the internet to communicate and forward their plans. The latest advancements in technology have always served as weapons in the hands of tyrants and those who wish to impose their opinions on others. My purpose is not to justify these horrible acts, nor is it to focus on the bad stuff. I'm not here to give only the optimists point-of-view. But I truly believe that focusing on the negative gives it more power and creates larger divides. I like to compare this to the automotive world; we choose to drive fast cars without the control of ensuring each individual driver is sober enough to drive. We also choose to share and collaborate on projects online, but we are not in control of the other users' links to terrorist groups or malicious hacking. Nearly 30 Americans are killed each day by drunk drivers. Does that mean you're not going to ever drive again? No. We should do everything in our control to reduce the drunk driver issue, but that doesn't mean we should all stop using cars. In the same way, we shouldn't avoid internet collaboration because a small (but vocal) percentage of the online population has been involved with malevolent hacks. So when you read through the book and see that I am praising the benefits of online sharing, do not frown or look at me the wrong way. I promise, it's all about striking the perfect balance and P2P can indeed achieve that. Join me in my pursuit to untangle the thread of P2P revolution!

About Me

Before moving forward, let me tell you a little bit about myself. I was born in 1988 in the U.S. of A, baby! California has always been my home and I am a first generation American whose mom and dad both immigrated from Denmark in the 70's. I am a small town tech nerd who loves reading. I'm just another millennial, who wandered around the internet in its early beginnings (1999 is when I first joined Napster at 11!) finding the wonderful world of P2P. Ever since I can remember, I have always loved sci-fi, tech culture, and futurism. At the same time, I have developed a keen eye on how modern technologies are engineered, as well as the role they are bound to play in our future lives. With a love for books and great storytelling, this will be my first attempt at sharing one of my stories with the world.

My life is deeply connected to music, as I have found it to be one of the few paths to collective consciousness that I have experienced in my life. Music in my childhood was life-changing, as it opened the doors for me to discover a new world of harmony and balance. In my teenage years, music invoked my revolutionary side, as it was the means for me to express exactly what I wanted to say. During my 20's and up to this current day, I use music to connect to my innermost feelings beyond language and symbols. Music never ceases to amaze me, and I can safely say it is much more than just sounds. It is indeed love—pure, unconditional love.

However, it has not always been easy to access the audio content of my liking. If you recall, back in my early days music was not free. So I had to restrict my urge to listen to various music genres and only stick to the artists I knew and whose CD's I had purchased. More than that, it would be next to impossible to experiment and eventually unveil the music that turned my world around. I would have had a much more superficial connection to the world of music if it weren't for P2P and the founding of Napster. Without it, I'd only know Top 40 hits and the Patsy Cline my mom blared every Sunday. Through revolutionary technology, I was able to get all of the music from every band I discovered along the way.

As a proud product of this wonderful era, I took an oath to promote the benefits of P2P in this book. Despite half-truths and misconceptions, open source information has helped me identify who I really am and I owe a great deal to the technological wonders of the internet. So let's start by painting a clear picture of why exchanging information online is an act of altruism and not of menace. This is the "matrix" moment, take the blue pill and the story ends here—you can stop reading. Or, be like Neo and let's see how deep this world goes.

P.S. In no way am I comparing myself to the great and wonderful Morpheous.

Chapter 1: The Wondrous Tale of Napster

Back in 1999, Shawn Fanning and Sean Parker had a vision about the way file sharing should be in the digital world. And remember, Napster is a "Black Swan," that is, something no one expected would change the whole process of online file distribution. The two teenage programmers did not hesitate for one moment. Not even when the entire music industry seemed to be fighting them with lawsuits and accusations. On the contrary, they held onto their dream and gifted humanity with one of the greatest technologies ever. Even its sworn rivals could not argue with the huge benefits derived from that amazing innovation.

The mastermind behind Napster is none other than Shawn Fanning, who spent his childhood summers working for his uncle's internet company. This was his inspiration for creating an easy and simple way to share files. To do so, he wrote code nonstop and even quit university to dedicate himself wholeheartedly in this endeavor. Finally, in 1999, he came up with Napster in its primary form. It was a P2P software program that enabled a network of connected computers to find music files and share them. As simple as that, Shawn succeeded in creating powerful computer networks that shared thousands of files online.

Shawn and Sean combined the exponential power of the P2P network with the use of reliable central servers.

This is why many people claim there are several predecessors of Napster to pay tribute to. For instance, IRC, Lycos, Hotline, and USENET are all networks that did pretty much the same thing. So why Napster was such a hit and why did it change the orbit of online file sharing? Mainly because of its exclusivity and its user-friendly interface. Napster was all about sharing music files. This meant you could find demos and uncensored versions, rare songs and performances at concerts from a plethora of artists. It was exclusive and featured hidden gems, as well as songs before their official release. Furthermore, it was easy to use and did not require any technical skills whatsoever.

The two friends joined forces and came up with a simple yet efficient business model. From the chatrooms, they gathered their workforce and eventually got adequate capital to move forward. With $50,000 in their pocket, they moved to California and set up their headquarters. It was a dream come true, even though Shawn and Sean could not have expected what was about to follow. The conditions were ideal for such a platform that connected the entire world in a single network.

With the increase of internet bandwidth, more people became familiar with the wonderful benefits of P2P file sharing. The numbers were mind-blowing! In 2000, 1 billion free music files were shared among Napster users (Pew Internet Project, 2000). Of course, this software entirely changed the way people listened to music. Now, there was no limitation related to money and resources among users. They simply needed to connect to the P2P network and search for the files they wanted to acquire. Depending on the speed of their internet connection, they could get the desired content and listen to it on the spot. I know this is a given in today's world, but back then this was a revelation. Nobody could have ever hoped to obtain a wealth of content within such a short time. So it came as a surprise and took the world by storm.

Napster Raises Ambiguity

Free file sharing, easy and simple for everyone to use. Who could fight against something so appealing? I am sure you have already guessed it. The music industry was not at all happy about the situation. Before Napster, it was next to impossible to get your hands on music for free. Instead, you had to buy the record or CD of the artist you wanted to listen to. A large share of the profit was given to the artists, while the companies also benefited through the purchase of music products. So the industry grew bigger and bigger. The prices at some point skyrocketed, since there was a monopoly in the industry. You could either buy the music you liked or rely on pure luck while listening on the radio.

With the increase of Napster's popularity, things began crumbling for the music industry. If you can get music for free, why on Earth will you pay for it? Whenever free is an option, it always prevails. As a result, a lot of artists saw a significant percentage of their earnings vanish into thin air. They felt threatened by the innovative software program that allowed everyone to connect and share music files without any charge. There were many complaints about copyright infringement. Unauthorized material was shared online and the music companies could not stand that. In fact, this was probably the end of an era for the music industry. People stopped buying music at the rate they did in the past.

On the other hand, Napster helped music spread across the globe. This meant artists could see their music become popular to a much broader audience worldwide. For example, the English rock band Radiohead had never had one of their songs climb at the top of Billboard until Napster came along. Then their album was downloaded by millions of people in the U.S. and got to number one right away (Wikipedia Contributors, 2019a)! If this doesn't paint the picture of how powerful Napster was in its peak, then what does? To use a quote that shows exactly how Napster developed a simple yet efficient business model and managed to keep it running for a while: "Napster created the internet's largest music library without owning or selling any music" (Gordon, 2005). If you are a fan of statistics, by 2000 about 14,000 songs were downloaded every single minute on the platform. That's a lot!

So even though many artists like Metallica or Dr. Dre opposed Napster as a means of music file sharing, there were others who greatly benefited from its use. This raised a huge dilemma as to whether or not copyrighted material should be shared online without authorization. Lawsuits came along and Napster had to settle more than once. The Recording Industry Association of America (RIAA) also sued Napster, with significant drops in the overall revenues of the industry. If you go by the book, it makes sense why an artist doesn't want to minimize profit. However, there are millions of people out there in pursuit of great music. Not all of them can afford to buy songs and albums on a regular basis. Should they be excluded from the music industry altogether?

As previously mentioned, there were some forms of file sharing before Napster. There is a thin line we all tend to forget or underestimate. Do you remember the time when cassettes were a thing? Besides records, you could buy a cassette and listen to music. Well, if you recall, you had to record individual songs onto a blank tape in order to create a "mixtape" or playlist. Usually you or a friend would create something like a " Summer '97" mixtape that was a dozen or so songs you either recorded from the radio or from other tapes you had already purchased. There definitely wasn't the police knocking at my door telling me to stop making mixtapes back in the mid-90's.

Getting Shut Down

Napster had undoubtedly been the world's largest digital music library to have ever existed. Its impact has been tremendous, bringing music into every home, every dorm, every campus—everywhere with very few limitations. Even years after its initial shutdown in 2001, Napster was the influential force behind a huge wave of music sharing programs released in the following years(*Peer-To-Peer Piracy On University Campuses*, 2003). Can you see how much Napster has affected broad and unconditional distribution of files online?

Until its conception and dramatic increase in popularity, people could not expect what was about to come their way. They had settled for a fraction of the music they would have liked to listen to. In a sense, they limited their desires. Before Shawn and Sean started this company, music fans had to face many limitations and restrictions when it came to their musical curiosities. Restrictions do not reflect the feeling of absolute freedom, which inspires artists and makes the audience ecstatic. For instance, many of the newer music genres are a mixture of older genres. Limiting access to music could prevent artists from developing new genres. And again, this has been the driving force behind the establishment of peer to peer software for file sharing online.

When the digital world was introduced to Napster, the shock was remarkable. People could not wait to be part of this revolutionary process that allowed them to grasp new music and explore entire artist discographies. Obviously, it is no wonder why Napster thrived on campuses and in educational institutes. This is where they had the most servers and highest quality internet. These were the sole requirements for using the software. If you did not have an internet connection, you could not connect to the P2P network and start downloading the stuff you wanted. Also, students have open minds and therefore willing to experiment with something new.

In the eye-opening documentary, *'Downloaded'*, there is a detailed analysis of the rise and fall of Napster, most likely the largest widely-accessible music collection the world had ever created at this point in time. *'Downloaded'* allows people to see both sides of the coin and eventually form their own opinion as to who is to blame. Gene Kan, the late P2P programming icon, commented on the shutdown: "What happened yesterday is the effective shutdown of Napster and what that's going to do is give birth to a thousand of Napster spawn and these children are going to be much better bred" (movieweb, 2012). Shawn Fanning and Sean Parker were the first to envision that file sharing could become real through a single network. You could connect to the digital world and take what you wanted in the form of MP3, as simple as that. That led to litigation, but nowadays things are much different.

After the owners liquidated their assets, the Napster name and logo were later used by other companies. Despite the efforts of creating a successor to Napster (in the form of Napster 2.0 and 3.0), the obstacles were too great. Who knows what the future of the "Napster" name will be, but we can all hope for the best. This company brought us two innovative entrepreneurs who significantly spread the use of open source software, and for that they deserve our thanks. The end of Napster in 2001 was just the beginning. After Napster, we got Limewire and KaZaA, which eventually reinvented itself into the modern day BitTorrent. Torrent sites multiply by the second, since they offer so much for so little. Well, if you look at it the right way, you will see that torrent sites are Napster's offsprings. Way to go, granddad!

A Moral Dilemma

Today, internet users who want to download TV series, music, films and books can do that through torrent sites. Of course, there are many complaints about torrent file sharing now. People criticize the free distribution of intellectual property and call those who engage in such actions "pirates." This is why torrent sites are often shut down, too. Napster was not the only P2P software program under scrutiny. Online piracy has been an enormous issue for the music industry, filmmakers, writers, and artists of all forms. A lot of money is at stake, raising a moral dilemma for everyone.

When something benefits the wider good so much, how can this be wrong? This is a question that often comes to mind. In the case of free file sharing, everybody reaps the benefits of technological progress and the internet serves as a vessel. Information is shared instantly, without compromising the quality of the file. As a result, more people can access a plethora of songs, films, and various other files over the web. The process is simple and straightforward. The only requirements include a PC, laptop, smartphone, or tablet, as well as an internet connection to support file sharing.

On the other hand, people have worked hard to produce their art. They have made sure to safeguard their property through copyright. Therefore, they demand they get the respective monetary gain out of their efforts. It is understandable people want to be rewarded for their work. Nevertheless, take a look at platforms like YouTube. They are also sources where anyone can access videos and music for free. Apart from the official channels that get paid according to the views of each video, there is no other profit out of the specific procedure. So in a way, the artists agree to expose their work online through these channels and offer people the opportunity to enjoy it for free. How is P2P file sharing so different and why is it condemned without any chance to redeem itself?

Napster was the driving force of the digital revolution for those who could not afford to buy music. High school and college students, the unemployed, and everyone else struggling with their finances could finally benefit from free music. At the same time, artists who sought after fame and exposure for their songs could do that through Napster. Demos and unedited performances, whole albums and even experimental songs could be found online so that the world was able to download and listen. As a result, with minimal marketing and promotional processes, a band or an artist could gain great popularity online.

Back then, when you had money and you decided to invest it in buying music, you typically bought a CD. One CD had 700 MB on average, which means it could fit an album. Sometimes, a single album was divided into two CDs. So you see that owning them was not easy. It required space and obviously it required money. Furthermore, these items were delicate and prone to damage. Even the tiniest scratch would take its toll on the audio quality. This was of course if you were lucky, because usually scratches and cuts meant that you had to kiss your CD goodbye.

With the introduction of MP3 as music files, the size for a single song shrank and you could have it stored on your hard drive for a mere 4 MB. The quality was decent, but not exceptional—and that is something you need to realize. Free distribution of files concentrates on the lack of premium fees. You do not pay anything and you can download music to listen to without any restrictions. Nobody would complain about the low quality of an audio file. If it were truly horrible, maybe they would rate it poorly to warn people and prevent them from making the same mistake. If it were unbearable, they would delete the file and search for another one. As simple as that!

You should not underestimate the power that came with Napster as a vessel of promoting one's work. A song was downloaded time and time again simply because someone wanted to listen to it. There was no promo available. So internet users could search and come up with rare editions, demos, and whole concerts. They used keywords to narrow their search results, just like we do while using Google. Someone would search for hip-hop tracks, while others would use Madonna or Elvis Presley as keywords. Some people would prefer to search for generic descriptions, such as relaxing music or best disco music of all time. But no matter what terms they used, there were always results waiting for them on the other side.

There is nothing wrong in recognizing the hard work of artists. But Napster did not aim at harming them, let alone messing with their monetary gain. Its founders thought outside the box and envisioned the future in a revolutionary manner, despite the fact that huge obstacles appeared along the way. This is the only path to greatness, after all. The concept was innovative and allowed *everyone* to enjoy music, no matter their social backgrounds or racial differences. Their goal was not to harm people. Instead, their vision was to help the community enjoy something they considered universal. They had to break the boundaries of music and the feeling was absolutely liberating.

What Next?

On the threshold of a new century, the internet went through the ceiling with catapulting speed rates providing a fertile ground for users to benefit from. Many apps and torrent sites emerged in an attempt to fill Napster's shoes and expand its features. Nowadays, it is possible to download and upload music and all kinds of files in fractions of a second. The conditions have changed radically and this is reflected on both the quality and quantity of the files distributed across the world. Internet users have become accustomed to sharing a substantial volume of content. Plus, there are websites and apps that allow you to subscribe and enjoy live streaming without glitches or problems.

Among the most widely recognized websites ever to have thrived on the internet is undoubtedly The Pirate Bay. It is self-described as the galaxy's most resilient

BitTorrent site. Established in 2003, it gained universal acclaim by taking into consideration the unquenchable thirst of people for films and videos. The Pirate Bay utilized P2P software to create networks of computers that shared the files. OK, it wasn't like visiting a video club and choosing which video to rent for the week. But again, it was free. There were some hidden gems there, apart from the latest blockbusters and films that had not been released at the cinemas. However, along with these files, you could risk getting your hands on some truly spooky files and I don't even mean it in the good way.

People have turned to torrent sites because it was a way out for them. Acquiring all those fancy books and albums, going to the cinema twice a week, and subscribing to premium TV portals is great. But how can you do that unless you have tons of extra cash to spend? There are people who cannot afford such purchases, but would love to enjoy some music and watch films online. Moreover, there are countries with heavy restrictions as to what is allowed and what is banned from streaming. What happens then? Should people compromise their own beliefs and only enjoy what is available and provided by the authorities? I do not think that this is a viable solution.

Social inequalities are mitigated through the use of open source file sharing, wouldn't you say so? No matter if you are rich or poor, young or old, educated or not, no matter your religion or any other artificial discrimination used to keep people divided, open source doesn't care. As long as you are connected to the web, you are free to access the files that are available in the network and pick what appeals to you. Certainly, they are not legal and hence many people hesitate to indulge in a file sharing spree. They feel guilty of exploiting the artists, but truly their work is praised rather than left to oblivion.

I can understand why torrent sites get all the rage. They fail to acquire the necessary licenses that copyrights dictate before distributing the content to people. But let's face it, if they did comply with the legislation and if they agreed to pay the fees required, then in what way would they be different from all other stores? What would be their unique selling point (USP)? No, P2P represents something mythical. It is much bigger than what we may think. It makes us connect with the rest of the world in anticipation of the next big thing online.

The digital revolution has taken place. There is a huge difference in people's mentality as to the internet and its use. The vast majority can now take pride in using these apps to reap the benefits of technology when it comes to entertainment. Consequently, owning content is not as essential as it used to be. In the past, it was necessary to download the music file you wanted first and then listen to it. Nowadays this is not a given. On the contrary, you can listen to music or watch movies through the use of live streaming. Why clutter your hard drive or store content on the cloud when it is so easily accessible for you at all times?

As time goes by and the internet is established as a fundamental way to connect people all over the world, solid legislation is needed. Laws must be written to protect both users and website owners alike. Since the web is relatively new, legislation needs to be created from scratch. But the time has come to get a foothold on internet rights. As a result, controversies will be resolved and the perfect balance will be struck between the users and the copyright holders of artistic content. In the end, these are not opposing sides. They all aim at the very same thing, which is none other than the evolution of *Art* in all its forms. Believe it or not, P2P is working towards that direction too.

Chapter 2: Following the Footsteps of Napster

Napster modified the way we listened to music forever, that's for sure. But open source is much more than that. It is true that many idealists have tried to change the world and the internet has been their driving force. Let's take WikiLeaks for example, which describes itself as, "An uncensorable system for untraceable mass document leaking" (Wikipedia Contributors, 2019c). This is a non-profit organization that exposes censored documents related to corruption in government and the corporate world. Most of the time, media coverage is biased and promotes several interests. In avoidance of biased information, WikiLeaks founder Julian Assange wanted to show the world what the truth is. Using open source and file sharing, WikiLeaks manages to provide uncensored material that optimize information distribution all over the world.

Since 2006, more than 10 million documents have been leaked and the repercussions have been great. All these years, the journalists working on this project have revealed huge scandals and shed light on the truth without any hesitation or hidden agenda. The freedom of the press is one of the fundamental rights of people. In totalitarian regimes and in cases of significant corruption, injustice and manipulation, information is crucial. Without the truth, you cannot expect people to defend their core values and strengthen their constitutional rights. Through a network of collaborated media and through the exposure that has been gained so far online, WikiLeaks can take pride in changing the world of information.

Up until very recently, the media did not have a lot to be proud of. Journalists would often be paid to distort the truth and present the world with a false reality about politicians and men of power. You would find them mingling with these people in galas and at the opera, in cocktail parties or aboard luxurious yachts. It was hard not to connect the dots and put two and two together. By all means, not all reporters were of the same quality. There were also those who could not be bought. Corruption was so deep, however, that they were isolated and could not really make a difference. In a newspaper filled with trash, how can a single moral journalist prevail? It seemed like a lost cause.

Media manipulation means people have little trust in their leaders. Some countries in the developing world, as well as European countries and the oligarchs of the Orient, made great use of the status in the media. In this way, they could promote exactly what they wanted. By feeding manipulated news to the public, it was inevitable they would accomplish their goal. Nonetheless, do not for one moment think that situation has remained the same for the press.

Instead of settling for media muzzling and preventing the truth from being accessible to the people, the news is now very different. There is always a beacon of light where everyone goes to for unbiased information gathering. WikiLeaks is not at all a "vigilante" haunting the dreams of villains, but this is a very eloquent comparison. Open source has played a key role in facilitating the distribution of important news, which has changed the way the truth comes out online—spontaneously, unconditionally, and for the entire world to see.

There have been allegations against organizations like WikiLeaks. Some people claim terrorism is promoted through these open source sharing options online, while others point out the fact that this kind of information might end up triggering a war. How can this be? WikiLeaks is just a vessel used to transfer information without boundaries. It is the right of everyone to access clean and unbiased news. Scandals ought to be exposed rather than covered up. Terrorism has nothing to do with this business model. Instead, organizations based on open source believe in 100% transparency and leave nothing shady to smear the end result.

When I say "whistleblower," you say? My money is you'll say Edward Snowden. Snowden managed to leak and publish a wide range of classified documents from the U.S. National Security Agency (NSA) in 2013 and proved to the world there is a global surveillance scheme going on (Wikipedia Contributors, 2019c). The world stood in awe as the U.S. government could not justify why it had been keeping detailed archives of phone calls, emails, and physical addresses. Just like in the movies, one man stood up tall and said "That's enough!" Snowden let the world see the NSA for everything they are, warts and all.

The U.S. government fought Snowden hard, filing charges of theft of government property. They even called him a terrorist. Oh, the irony! Authorities have been caught red handed exploiting the governmental infrastructure to promote their own interests. They have been spying on people and they have been doing so for years. In fact, chances are that they never stopped, not even after the revelations from this whistleblower. And even under these circumstances, they still had the nerve to file lawsuits and claim he breached his security agreement, since he had signed a nondisclosure contract with the U.S. To put it bluntly, it is like a wannabe murderer attacks and nearly kills me. I fight back and manage to injure him in the head. Time passes and months later I receive a letter, informing me that the criminal is suing me for his injuries. Oh, the irony!

Edward Snowden did not go gently into the night, though. He had accumulated up to 200,000 NSA documents and had a strong bargaining chip. As to what has happened so far, it is safe to say that Snowden will not be coming home anytime soon. He sought asylum in several countries and finally settled in Russia. So now he continues on his path to disclosing top secret documents, with the aim of exposing the size of the surveillance operation. To some a patriot and a hero, to others a traitor who debunked the myth of the U.S., he is definitely an intriguing personality and is now an advocate for free online information.

More Than WikiLeaks

Do you remember the Panama Papers? Their disclosure rocked the global corporate world to its foundations in 2016. A whistleblower called John Doe (touché) leaked a massive volume of documents from Mossack Fonseca & Co. revealing fraud, tax havens, and various other financial crimes. Well-esteemed celebrities, politicians, and other powerful people were found to have some sort of relation to this law firm from Panama, which naturally crumbled and vanished into thin air. Corruption was found in literally every single part of the company, while the world was shocked to learn about deals involving former British Prime Minister David Cameron and Iceland Prime Minister Sigmundur Davíð Gunnlaugsson. And this was just the tip of the iceberg.

Even such groundbreaking news and revelations did not lead to World War III, as many thought. There have been people ready to condemn that enormous leak of documents, rather than focus on the real culprit. However, leaks have led to changes in the universal scene of politics and shed more light on law firms dealing with tax havens overseas and international banking. There is nothing wrong in knowing what is going on around you. Ignorance is not bliss, no matter what you may have heard in the past. The more you learn about the world, the better you prepare yourself to cope with its challenges.

You all know the group Anonymous. Having inspired millions of people all around the globe, this movement originated in 2003 and has been going strong until now. Wearing Guy Fawkes masks, people have united behind a common cause. Their goal is to bring light to the unknown scandals of governments and businesses, unveiling corruption and illegal behavior. Their members are tech-savvy—mostly hackers—and people with a deep sense of understanding the quintessence of solidarity who have joined forces. Although they swear by the principles of anarchy, they have a code they never break. Their aim is not to seek avenge for crimes committed, but solely to expose those crimes and inform the public. This is the best revenge, after all. So violence is not embraced. In fact, it is avoided at all costs. People are warned to stay vigilant, up-to-date with what is going on.

Do you want to know how much the world really needs a hero? Someone to relate to, someone to look out for their own interests and someone who is going to criticize everything that is wrong with society, aiming at creating full transparency? Well, you don't have to go very far. Simply check out *Money Heist*, the most popular series in Spain and other European countries broadcast on Netflix and you have your answer. *Money Heist* (Pina, 2017) has managed to earn universal acclaim and the plot has everything to do with it. The Spanish drama series grew an exceptional fan club right away, along with the song that became a hit on the spot.

As for the plot, it is quite intriguing and promising till the end. A group of misfits are guided by "the Professor" in their attempt to rob the Royal Mint of Spain. The plan seems to be impossible. The building is heavily guarded and it has never been breached before. However, the heroes of the Revolution (this is what the Professor calls them) are dressed in red and wear the masks of Dali—you see the resemblance with the Anonymous group, right? It is uncanny! Are they going to make it or will the police end their dreams sooner than expected? I know that all the fans are rooting for them, opposed to the strict government and the even stricter law enforcement.

I am not going to spoil the ending for you, but there are many twists and turns for you to watch for. As for the public response? Well, suffice to say that there have been several protests around the world with people dressed like the heroes of the Revolution. There is a traditional Italian folk song called "Bella Ciao" used in the series. Try not listening to that for one day! You will get hooked, it's that catchy. A couple even used it as their wedding song, I kid you not!

Again, I am not using any spoilers—enough said. Don't you notice a pattern in these scenarios? It is not far-fetched to assume that there is a growing part of society in search of this non realistic world. Is it so bad to dream of a place where no one suffers due to their unbearable debt? Is it so hard for communities to understand that the only way to push forward is to stick together and join forces? As it turns out, many people feel isolated. They are alone, voiceless and without the ability to stand up for what they believe and protest. Why not end their suffering by introducing the new digital world?

Online information doesn't ignite violence by default. How could it? For the vast majority of people, online information is absolutely harmless. There are those of course who are prone to delinquent behavior. They might end up using it for the wrong reasons. But this percentage is way too small to define the entire society.

Think of it this way: a traumatic experience may result in psychosis or lunacy. However, does this mean we should hospitalize all those who have been traumatized? Acts of terrorism or assault are orchestrated by people with no moral code. Online information is not to blame. If anything, raising awareness against such delinquents can be achieved online. The web can be used as a means to spread the word more quickly and efficiently. Isn't this a good thing?

Hoarding Information

Have you ever heard of the saying "less is more?" Don't get me wrong, information should never be censored or kept from the public eye. Nonetheless, it's important to differentiate between what's valuable information and what's valuable about having access to all the information. On the one hand, it's all about using and sharing files you value. On the other, it's about not overwhelming yourself and parsing out what is actually useful.

"If more information was the answer, then we would all be billionaires with perfect abs," according to the American writer Derek Sivers (Ferriss, 2015). This is so true. If only things were that simple. However, it requires critical thinking and evaluation of the information received. Your mind is like a filter, so you should use it as such. Identify what is indeed valuable and worth gathering and reject all the rest. Otherwise, you will soon find yourself overwhelmed by information and unable to handle what is coming your way. Imagine stocking up on unnecessary gossip and things that you didn't even want to hear about in the first place. And now think of the capacity of your mind. No matter how much you try to stretch it out, things will eventually spill over once you've hit your max capacity; the brain only has so much bandwidth to hold and remember the day's information. Mental acuity means knowing what not to think about as much as what to think about.

American author Chuck Palahniuk describes the situation most eloquently in his novel *Adjustment Day*. "Hoard food and it rots. Hoard money and you rot. Hoard power and the nation rots." (Green, 2019). The same goes for information, obviously. There is no point in hoarding anything in life, let alone information. You have now access to an unlimited volume of data, thanks to the internet. At any given time, you can search and find thousands of relevant sources of information to enlighten you. So what is the reason behind all this obsessive hoarding if not greed? By all means, I should not be the one to tell you how to live your life. Still, a friendly piece of advice. Ask yourself twice whether or not you can live without that one thing you've been keeping away from the eyes of others. Who knows? Maybe you'll find it liberating with fewer strings attached!

All this time we have learned to appreciate our personal belongings and sometimes we even go to extremes, defining our self-worth based on these very possessions. How can this be the case? Are you only as good as your new iPhone? And what happens when you get stripped of your Gucci outfits? I am sure there is much more than what meets the eye. So why do you keep belittling yourself, thinking that you should always buy more and more stuff to feel complete? In this era where materialism prevails in all aspects of our lives, it can be hard to reassess your priorities. But it will also lift an enormous burden that has been holding you down all this time.

Hoarding has invaded our modern world and has enslaved us, making us always strive for more. This has got to stop. By focusing on what is really important and by seeing hoarding for what it actually is, we can reverse the situation. It is true that it won't happen overnight. In fact, it is a slow process to change your view on materialism. This book isn't trying to get you to throw away all your possessions and start living some sort of minimalist lifestyle. It is important though to point out the parallels between people who stow away information for their possession and people who strive to own a lot of "things;" these are damaging practices in the long run and disconnect you from your community.

Another thing to keep in mind is the fact that fake news thrives through hoarding. Propaganda, hoaxes, biased misinformation, how can you protect yourself, unless you establish critical thinking? It makes perfect sense that you devote some time and effort towards evaluating your sources. By sharing what you have seen on some accessible database, you allow others who are interested to cross reference what they believe and update you on it's validity. Fake news requires people who rely on limited sources, much like those people who believe wearing Gucci makes them wealthy. Yes, there's a small percentage of the population who actually believe this to be true. Instead, read to satisfy your curiosity and always try to cross-check the facts. The spreading of false information online is often driven by impulse reactions from buzzfeed titles that catch attention and ignite emotion. But when it comes to real information, try to consider the source and not let your emotional self make rash decisions.

Fake news websites are like the plague. They belittle the value of information and only focus on volume, because this is their only chance to prevail. They flood you with more information than you could possibly handle and they hope some of that will be accepted without any critical eye raising concerns about their credibility. If you react, they will simply move on to the next fake news update and start over again. Obviously, one of the best ways to truly realize if an article is fake is to check the source. Meaning, take a look at the website you have navigated to. Does it seem legit? Is it the website of a well-esteemed newspaper or local news organization, or maybe the website of an aspiring young journalist? Or does it look shady? Is there any red flag that keeps bothering you? When in doubt, just run!

How to Enhance Your Online Security

Up until now, I have been showing you that open source file sharing is nowhere near the demon you might have thought it to be. Napster, Kazaa, Bittorrent, and WikiLeaks, all these have one thing in common. They were all born out of the need to share online and keep nothing in the dark. That being said, there is indeed solid ground to the accusations regarding online hazards. The web is not a playground and therefore you have to step up your game, if you want to stay safe online. How can you expect others to care when you do not lift a finger to enhance your online security? Excuse me, but this is just plain wishful thinking. You know better than this, so take matters into your own hands and let's see how you can shield your online experience!

- First and foremost, be mindful of what you share online. Check your privacy settings on social media. You will be surprised to see that friends of your friends can see what you are up to on Facebook, even if you ignore their very existence. How spooky is that? Plus, please don't share pictures of your children or anybody else's. Oversharing information with pictures is very easy. Photos of a birthday party or a vacation are fine. But photos of the park you go to every week, or a subtle caption that reveals when you're away from the house or that you go to the same family resort every year could be dangerous. Your phone number or physical address should also be kept private, unless of course you are a professional in pursuit of expanding your business network. Even then, use your business info rather than your personal data.
- Strengthen your passwords. It is highly likely you have been doing that wrong all this time. Sure,

there are many websites and apps encouraging you to use powerful passwords that are very hard to guess. Still, instinctively you might turn to techniques that make it easier for you to remember those passwords. As I told you earlier in this book, there is no point in your remembering all that information! Just use a password manager and you will be fine. Renew the passwords every now and then (add a reminder on Google Calendar) and *voilà!*

- Another online hazard takes place with your own permission. You authorize the penetration of your privacy even if you do not quite realize it. The truth is that it happens in a sophisticated, subtle way. You want to install an app on your mobile phone and you go to the market, hit the installation button, and then read carefully through the permissions. Yes, I know that nobody reads through them—but they should! This is how the apps ask for permissions that are relevant to collecting information for marketing purposes, etc. Similarly, browser extensions often come with permissions to be used for marketing. It is in your hands to change that.
- Even though Dropbox might seem appealing and Google Drive has been revolutionary, they are still not 100% private. Would you ever leave your jewelry in a locker that doesn't even have a key? So you should never consider storing sensitive documents or any other file that you want kept private in public storage options. The only exception would be if you encrypted the files first. After all, you do know that Dropbox, etc. are

mainly focused on sharing files, not storing them neatly away from harm.

- It is essential that you avoid online tracking. Every time you visit a website, there are breadcrumbs that you leave behind. These bits and pieces are widely used by third parties so they can better comprehend your likes and preferences. As a result, you get personalized ads that may seem random but aren't. Now, some people claim that Chrome's Incognito web browser is the solution to this problem. Although it does make a difference, it does not let you surf the web completely carefree. In order to do so, you must first choose the right tools for ensuring that you leave no trace behind.
- Online communication is great, as long as you understand its limitations and work around its flaws. For example, unencrypted communication might jeopardize your personal information and you do not want that, do you? If you want to share sensitive data (like your passwords, credit card credentials, etc.) and there is no other way than connecting online, then select an encrypted messaging app. WhatsApp is a wonderful option, with end-to-end encryption. On the contrary, Facebook Messenger does not have the same level of encryption by default. Luckily, you can tweak their settings and get that extra level of security.
- Moving on, it is important to go the extra mile for improving your online security. Even if you feel this is too much, it is probably still not enough. So if your mobile phone or your laptop comes

with an option to add a password, then do that. More than that, install a tracking app that allows you to monitor your device in case it gets stolen. Find out more about two-factor authentication and make it a habit. Prior to connecting to your mail, you will need to type a password and get a code on your phone. Do you remember a little earlier when I told you that sometimes "less is more?" Well, this is not one of those times!

- Finally, I would like to talk about Wi-Fi networks. Oh, those hotspots have saved my life more than once! They offer you the opportunity to connect to the web, even when you have run out of data or your signal is too weak. So nobody is going to argue as to how useful they can be. But you should be extra careful when you use these public Wi-Fi networks. Otherwise, you may end up getting scammed. Such connections are filled with malware and other threats, spying on your every move and eventually tracking your private information down like prey. One fine solution would be for you to enable VPN encryption, keeping all those dangers at bay.

It is equally important to make sure you avoid several cheap mistakes that might jeopardize your online security. Newbies are naturally more prone to falling for these tricks, but it is easy to let your guard down even for a second and become vulnerable. It is obvious that you should combine the aforementioned instructions as to how to shield your online presence, along with these "no-no" actions.

- One of the basic rookie mistakes you can make when it comes to the internet is to use out of date software and apps. You may think updating only affects the performance and speed of your computer. However, this is not the case. Software that does not feature the latest advancements is susceptible to threats and is most vulnerable. In fact, some of the upgrades may be relevant to dealing with those threats. So you need to avoid obsolete software like the plague. Invest some time to download the current version of your software, as well as all the tools you use when surfing the web.
- Some people are falsely convinced installing antivirus software is enough to protect themselves from online threats. Well, I wish things were that simple! Sadly, you need to do more than that. In fact, one solid piece of advice would be to engage in regular security scans. Even if no threat pops out, you will most likely notice several eye opening things, like which apps consume your disk's capacity, etc.
- Clickbait is always creeping around us like a wolf in sheep's clothing. Most of the time, it comes in the form of shocking news updates and articles that you are invited to click on for learning more. Alternatively, some tweets or even newsfeeds include shortened URLs that redirect you to another website. Now, it is not always shady, but this is a once in a lifetime opportunity for the marketer to manipulate you. Do you think you are that strong? I hope so, for the sake of your

online security. However, check twice before clicking on any link to see if the site is credible.

- Spam and junk email folders are there for a reason. And maybe you consider your spam filter is way too strict and prevents you from seeing emails you would actually enjoy reading. Still, remember you should never open emails from strangers that appear to have been created by bots. When something doesn't feel right, do not weigh your odds. Why take chances and risk your online security for whatever "deal" or "offer" the email claims to have? Besides, most email providers detect phishy email senders and this is why they place them in different folders—so you can stay safe online.
- When you visit websites that give you a feeling of insecurity, trust your instinct. Unfortunately, there are many websites out there that should be banned. During their short life span, however, they manage to make our online experience less appealing and safe. If you notice too many typos, obsolete themes, or slow page loading, it is time to hit the road. Instead, aim at visiting websites that have been tested and found credible and reliable. Of course, Incognito might also help you avoid such issues generated by shady websites. But its protection is not complete, nor should you rely just on its ability to mask your personal info.
- You know better than to leave all your money in one place or invest on a single stock, right? Then why aren't you prudent enough to backup your files and ensure continued smooth functioning, even in the event of your PC crashes? It is the

same as buying a house and keeping it safe by finding the right insurance that puts your mind at ease. Although you do not think a tsunami is possible in the mountains, there is no harm in checking that box at the insurance features.

- In case of an emergency, it is always good to know you have the things you can control under control. As a result, you will feel less stressed and this will help you to encounter any challenge in the most efficient manner. Well, as far as your online security is concerned, you may want to enable remote access and monitoring. In this way, you will not be in the dark. You will have the chance to run diagnostics and discover if there are any threats that need to be addressed. Why not even act the role of a vigorous detective trying to unveil mysteries? You can do that on your mobile phone or on the laptop, whatever serves you better.
- Finally, one word of caution. It has been proven that one of the most common sources of viruses and threats for computers is the USB flash drive. The reason behind it is pretty clear. Because of its compact size, the USB is widely used by people to exchange files and documents. It is only fair to assume not all computers are safe to use. Unfortunately, the minute the USB is inserted in your computer port, it activates. Any virus that has infected the original computer will most probably be passed on to the flash drive. You guessed it, the USB will in turn have increased possibilities of infecting the next computer it is used on and the cycle is repeated. So, be

meticulous when using a USB. Keep it clean and always scan it for viruses.

These are tips on what you should and shouldn't do. Think of them as the Ten Commandments (actually, they are far more than ten) of online security. Adhere to these instructions, as they can make your life so much easier. Don't ignore the warnings. Sometimes we choose to underestimate just how serious a situation is. As a result, we end up regretting our idleness. It's better being safe than sorry, right? So don't be lazy when it comes to safeguarding your online presence. Educate yourself on online security measures and be sure to implement them customized to fit your own needs.

Don't get me wrong, as a standard a Zen mindset is absolutely fabulous. It allows you to get rid of all the tension, the anxiety, and the stress, breathing in and out slowly and appreciating the moment. Sorry, I got carried away visualizing the perfect hot yoga session. Anyway, my point is I am not advising you to panic and go overboard. I am simply saying you need to be careful and consistent, security-wise.

What do you think? Are you ready to take your online security to the next level? Just make sure you follow the guidelines above and you will be great. Why would you fall for the oldest trick in the book and then blame the book itself? Gather all the information you need and shield your online presence. In this way, you will be less stressed about potential threats that are lurking in the darkest corners of the web. So you will have more time and energy to enjoy all the good things the internet has to offer. A win-win situation, for sure!

A Hoarding History

Let's travel back in time and land in 14th century Bavaria. Joseph Ritter von Fraunhofer was a local physicist and optical lens manufacturer. He spent his life trying to perfect his scientific discoveries. Fraunhofer developed a technique to polish large surfaces of spherical glass objects, inventing several grinding and polishing machines for that purpose. He also discovered how to make flint glass—that is, glass with a high refractive index—which led to his invention of the spectroscope as well as development of many optical instruments and telescopes. An amazing start up kit for astronomy, right?

One would expect that such a revolutionary invention would soon be spread all over the world. After all, mankind would definitely benefit from such knowledge and therefore it was some sort of sacred duty on behalf of the Bavarian government to disclose the Fraunhofer's inventions. However, this is not how history played. For over a century, these magnificent discoveries were kept a secret of the state. As a consequence, humanity's understanding of the universe remained idle for decades. Think of it this way. That Bavarian government was presented with the opportunity to open up new frontiers of space. Yesterday, you were in the dark and now you are given the opportunity to further understand our universe. Wouldn't you be mad if someone held onto those tools for over 100 years?

American astrophysicist Neil Degrass Tyson described this shocking fact in his documentary television series *Cosmos*. "The government kept Fraunhofer's technique for making perfect optical glass a state secret for another 100 years after his death. This would prove a major obstacle for [scientific advancement]." Fraunhofer's work would lead to the proof of "the stellar parallax." For the first time, the stellar parallax would allow humankind to track and chart the movement of celestial bodies. This would ultimately prove heliocentrism to be a scientific fact. Joseph Ritter von Fraunhofer died in 1826 at the age of 39. Yet the Bavarian government held his research and methods as a state secret; it wasn't until after 1901, when Annie Jump Cannon proposed using the "Fraunhofer lines" as the basis to create the system of star classification that we still use to this day! The hoarding of these tools went against what Fraunhofer stood for (as well as what this book stands for). "Fraunhofer would allow no such secrecy whereas scientific research was concerned. He knew that science requires openness to flourish and that our understanding of nature belongs to the world" (Braga et al., 2014). A true discoverer, a pioneer, a man of exceptional ethics with a vision that focused on what was good for the world, Fraunhofer would never have reacted the way his government did. This was a form of betrayal on behalf of his kin, as they did not respect his life's work and what he wanted to achieve.

This brings us to a sad acknowledgement. Sometimes the true heroes of our society are treated wrongfully by the authorities. It has happened again and again, often with harsh treatment. Galileo was persecuted by the Catholic Church for suggesting the earth revolved around the sun. But he wasn't alone. The great Greek philosopher, Socrates, had a wonderfully bright mind and developed the Socratic method (dialectic method of inquiry) which is essential for scientific reasoning. Socrates said we ought to question everything in life. Nothing should be taken for granted. Furthermore, he claimed we have the truth deep inside us and it is our duty to let it out, through the proper questioning.

Socrates did not let anyone escape his criticism, if he felt they deserved it. He was no saint, but he wasn't biased. However, his arrogance and strict character led to sworn enemies in Athens. The authorities put Socrates on trial, during which they would determine if he was guilty of corrupting the Greek youth and showing disrespect to the gods worshiped in Athens. Obviously, Socrates knew this was going to be a complete parody of a trial. Yet, he endured the entire process so as not to mock democracy. Socrates was sentenced to death by poisoning and forced to drink hemlock. Although his students begged him to go away and escape the death sentence, he complied by the rules to show his unquestionable respect to the idea of justice.

Let's return to the story about Joseph Ritter von Fraunhofer. When you keep such earth-shattering discoveries as his secret, in a way you plot against humanity. An understanding of nature belongs to the world, so such knowledge is not yours to hide. In the same way, the world of the internet should be accessible to everybody as our cultural legacy. It is part of who we are and it shows our constant evolution. Drawing a veil over the web and creating a false understanding of truth is a crime. Information and knowledge should be out there, shared with the world through open source. Far away from the obscurantism of the past, the internet reflects the concepts of freedom, equality, solidarity, and unity. How can this be wrong, I wonder?

Chapter 3: An Online Philosophy That Works

Let's start this chapter by visualizing the perfect day out shopping. I know the rebellious part of you will now be mocking me for my choice, but please bear with me. You are in the mall or wherever you feel most comfortable, maybe down a boutique retail avenue. You stroll up and down the aisles, searching for the things that tickle your fancy. Well, most likely you have a specific budget and you cannot go over it, no matter how much you love that silk scarf or the gadget that would make your life at the office so much easier. Either way, you walk slowly past each store and indulge in some relaxing window shopping. You notice what is on display, often wondering if they are indeed worth the money or if they are ridiculously overpriced.

Suddenly, an astonishing display catches your eye. It is a sight for sore eyes, I give you that! The perfect outfit that will make everyone look at you with envy. You would be a femme fatale or a Beau Brummell in that, enticing those around you to talk to you and come closer. This would be the most powerful item in your arsenal, impressing clients and associates, coworkers and rivals. Amazing attention to detail, sophisticated garments that seem to have been created with you in mind. Normally, you would storm into the store and buy that outfit in a heartbeat. Still, you glance at the price tag and you immediately feel your two feet back on the ground. Then again, is it you or is that outfit calling your name?

You have two choices outlined in front of you and it is in your hands to decide which way to go. The first would be to spend some time admiring the clothes that are inviting you to buy them. Look at them closely, see how elegant and posh they are. Imagine wearing them, where exactly would you go and what exactly would you do? Then, after having enjoyed some moments of daydreaming, move along in pursuit of something that does not require several body parts to be sold to pay for them. Rationalize the idea that it is just an outfit and it does not affect your quality of life. You do not need it, right? It is not a matter of life or death, so why bury yourself in debt over it?

The other choice would be to give in to temptation. You go in and buy that stunning masterpiece of contemporary art. You deserve it! Besides, money is only good when you use it to make yourself happy. Buying that outfit has definitely made your jaws hurt a little from smiling so widely. Perhaps you have already started thinking of ways to cut down on your other expenses, so that this purchase does not destroy you financially. Either way, you can't wait to wear it and make an entrance for the rest of the world to see! So bottom line, you are cool with your decision. Sure, a sudden impulse may have you running back to the store for a refund, but that's life!

I don't know which version you empathize with most, but one thing is certain. Most people would react in either one of these ways, eventually standing by their decision and moving on with their life. This happens 9.9 times out of 10. Sure, there is always that tiny percentage that messes with the world's balance, let alone with your head. In this case, that mere 0.1 would sneak the suit into their backpack and run off with it, or plot some eloquent plan to come back after hours and steal it without anyone seeing. This type of person will always exist even though they are a small percentage of the population. These aggressive, criminal types could care less for the order of society and choose thievery as their main course of action. Anyway, they would barge in and smash the window display, act aggressively and terrorize the staff in an attempt to validate their dominance and get what they want.

They would spend some time admiring their latest possession and after that moment has passed, they would most probably start figuring out ways to sell it to the highest bidder. It goes without saying that this 0.1 needs help. They often come from broken homes or deal with hardships that are tearing them apart. Alternatively, they suffer from severe mental states and do not receive the proper treatment. As you can imagine, this is not an idyllic way to live. Their actions are not justified, but it is always good to dig deeper and try to find the root cause of the problem. Once that problem is gone, that 0.1 will channel their reactions to one of the two options described earlier.

It is the same with information found on the web. Just because information is available and is displayed for all those passing by to view, this doesn't mean everyone will use it to exploit others. Anarchy would prevail, if that were the case. People want everything to be available to them and within their reach at all times, but they will only pay attention to things they are interested in. Otherwise, they will not even bother. So the abundance of information shared online is accessible to all, even though not all end up accessing it.

The internet is not a place where we need to be in a fight or flight mode at all times. In many aspects, the internet reflects our society. There are not many differences, to begin with. The percentage of malevolence online is proportionate to the percentage you would encounter by walking through your neighborhood. When you are walking down your street, do you feel safe? Sure, there is always the primal instinct to survive, the need to protect yourself. Still, walking around your hometown is not the same as landing in the jungle and having to survive with nothing but your two bare hands. Well, the internet resembles society more than it does the jungle.

Of course, we want to decrease that percentage of malevolence online even more. In fact, it would be great to eliminate it altogether (even though this sounds like a utopia). To do so, detailed legislation has been established that allows internet users to surf the web safely. It is a work in progress, with ongoing improvements and additions to the current laws. Though still evolving, the field of online security has made the web a better place. Nowadays, every country has dedicated law enforcement departments that deal exclusively with the internet and all things relevant to the web. This is to ensure people sleep better at night, knowing that someone is watching out for them. Even if the entire concept of law enforcement might be deterrent for some to begin with, it is unquestionable that such protection is needed.

From Fight or Flight to Share and Surf

The internet is a magnificent place, where all the magic of the digital world is unveiled before your eyes. Yet there are people who manage to use this magic to carry out the evil deeds. If you let it, the internet can take you into some pretty dark places. Does this mean you need to pick sides? Are you for or against the internet? No way, things do not work like that online. Obviously, the biggest fear regarding the web on a global scale would be to harbor criminals, mostly terrorists. Terrorism is a broad term that covers fifty shades of evil, right? So after a terrorist act that has cost the lives of innocent human civilians, it is not unusual for people to doubt the value of the internet. Is it really worth it? When all those kids are put at risk, when whole countries can be destroyed in a blink of an eye, how can the internet be allowed to nurture this evil?

Terrorism is done with a political agenda. So this means many databases and information sectors of the internet have no use to these groups. It would be crazy to assume that a terrorist would benefit greatly from free file sharing of music in the MP3 format. They use the web as a vessel in order to communicate with each other and plan their notorious schemes. OK, this I can understand. However, before the internet, there used to be landlines. This was another means for terrorists to communicate. Is it prudent to ban the use of landlines as a whole? And what about pen and paper, the post office, or even the TV? When you open a can of worms, all hell breaks loose and it is next to impossible to restrain that.

Think of it this way. Yes, an innocent foreign civilian can be captured and held hostage for a terrorist group's agenda. This has happened before in the past, so there is no point in arguing about it. But how probable is that to happen? Have you ever met a person, who was kept in captivity by terrorists? I bet you haven't. The vast majority of people traveling around the world will not be kidnapped, nor will they live anything near as disturbing. More than that, their life will truly be enriched through the whole experience of discovering new places and new cultures. Traveling broadens your horizons and should not be avoided for fear of what *might* happen. So why spoil a life's dreams and hopes because of something that should not even make it to your list of worries? Deciding not to travel due to the negative feeling of discomfort that you are experiencing is understandable up to a point. However, at the end of the day, it is irrational to avoid world travel simply because of a fear of the possibility of terrorism.

Now take a moment and think of your online information. It is indeed available for everyone to see, even though as mentioned above you can enhance your online security and pretty much control who you expose your personal details to. But let's assume for a moment you do not care a lot about preventive measures. Even then, having your information online doesn't mean you are going to get hacked! Will you be constantly stressing about the possibility your data gets intercepted and used as leverage against you? How probable does that sound? Let's be real, this is not highly likely. To be honest, it is the same as if you traveled to a country like Argentina, Indonesia, or Pakistan; do you truly believe there is a high probability you'd get kidnapped in one of those places? OK, you can roll the dice and see what happens. Either way, the odds are in your favor.

My goal is to set you at ease. It is counter productive to constantly feel like that, so don't! If you are concerned about your online information, then do something about it. Make sure that you use up-to-date apps and tools for strengthening your defensive system against hackers and malevolent acts. You are in control of the level of transparency you get on the web. When traveling, there are things that you can do to mitigate unnecessary threats, isn't that right? You should avoid walking alone at night, steer clear of dodgy neighborhoods, stay in groups rather than stroll alone, and never make it easy for anyone to exploit you. Why keep the wallet in your hand, why leave your mobile phone unattended, why get that map out to let everyone know that you are a tourist?

In the same way, there are always additional measures you can take towards improving your overall online security. This does not happen overnight, so be sure you are consistent in your efforts. Do not give up too soon or you will not have the chance to reap the benefits of your prudence. And one last thing: the fact it is extremely improbable that you will get hacked does not necessarily mean you should rest assured it will never happen. I am not saying you will all get hacked or hackers will try to penetrate your online security. But you should never think it's impossible because this will only weaken your security in the long run. This is the main reason why it is essential you comply with the guidelines mentioned above. On top of that, this is why cyber security is of utmost importance for keeping the web a safe place to be.

The Importance of Cyber Security

As I have probably hinted more than once so far, I am not advocating for the "defunding the internet police." Cyber security is a very real thing and everyone should do their part to keep passwords, home and work addresses, and personal information secure. Furthermore, people should get over themselves and not hand out such sensitive information all willy-nilly like. There are professionals who dedicate their life towards making the web secure. They think outside the box, trying to foresee future threats that will flood the internet. In this way, they have been working hard to nip those threats in the bud. Law enforcement has also evolved a great deal, devoting both manpower and resources for the ongoing battle against hackers and online threats. And their efforts have paid out, too. There should also be a well deserved shout out to all those ethical hackers out there. They make a habit out of hacking into computer systems simply to point out their vulnerabilities and contribute to their security optimization. Isn't that lovely?

As technology progresses and cyber security analysts come up with solutions to encounter threats on the web, the threats have also become more difficult to handle. After all, hackers could do nothing but to fight back. They upped their game and formed groups to put their notorious skill sets to use so they can pose new threats to the digital world. So this is a continuous battle and the results are not predetermined. It takes a lot of time, hard work and even luck to predict the outcome. Nonetheless, even when they win, hackers will find it very difficult to crack encrypted code and penetrate secured, sensitive documents. So, I repeat—it is very important to know what not to post online as well as how to protect your valuables that are online.

Let's talk about celebrities for a while. They are always in the spotlight, exposing themselves to the public. This is part of their job description, after all. Due to the nature of their work, they develop a personal relationship with their fans. To do that, they make use of social media platforms. Selfies, stories, giveaways, snapshots of their everyday life are used to communicate with their audience. However, this comes at a significant cost on their part. They risk revealing too much about their private, personal life to complete strangers. Doesn't that sound frightening?

By letting their audience peek through the keyhole on a regular basis, they provoke a feeling of intimacy. In most cases, this feeling is harmless. Nevertheless, it can get spooky at times. People who have become obsessed with celebrities can attempt to penetrate their privacy. If online security breaches were easy to accomplish, then we would have been looking at reality from a different standpoint. Luckily, it is nearly impossible to suffer from identity theft and data breach, as long as you comply with the above mentioned guidelines.

Before wrapping up this chapter, I wanted to point out a slightly different aspect of online experience. People love the narrative of other people's life stories. This is a must, when you want to become an influencer. However, that feeling of looking through a window and gazing at a person's life is quite therapeutic to some. It helps them forget their own life, their pain and suffering for a while. Instead, they get the chance to observe from a distance. They analyze and form solid conclusions. This is a whole process, a coping mechanism for society that should not be ignored.

So it is not rare for people to form connections online that often mimic the traits of friendship in real life. I am sure you know what I am talking about. Besides, on the internet you get the opportunity to mingle with a rather unusual group of people. There are no restrictions as to age or sex, origin or background, profession or any other things that may have kept you apart in real life. You begin talking about something and you end up sharing secrets you have never shared with anyone else in your life. It is a combination of cathartic talk with someone who does not quite know you and hence cannot be judgmental, as well as the excitement that derives from a new acquaintance. Online there are distorting mirrors that you can use in case you do not want others to see who you really are. I am not referring only to the way you look, but also to your character. You can choose who you want to be, or reveal your true self—always with respect to online security measures.

The internet opens up a whole new world of opportunities, as far as meeting new people is concerned. On top of the massive quantity, you also enjoy an amazing quality of connections. You can find your soulmate, romantically speaking, or come across your next business partner. There are tons of people out there, waiting to meet you. Depending on your likes and preferences, your habits and online navigation patterns, you can meet exciting new people who are like-minded and share the same passions as you. Then, it is only fair you evaluate your current status and start looking at life in a whole new perspective.

And this raises the following question. Why be friends with the people you have grown up with simply because of proximity or comfort? Is it something you feel obliged to do? I don't think such geo-restrictions make much sense nowadays. In the past, things were different. If you listen to your folks talk, they will most likely reminisce about the time they got to spend with their friends, doing something gross like collecting worms or eating their snot. It is all wrapped in a festive description, of course, which appears to make even grown men shed a tear or two. Times were much harsher back then, but we could still taste real tomatoes and cucumbers and squash. I guess nature compensates the poor for all the bad things they have been through.

In all fairness, you will grow old (I certainly hope so) and you will shed tears when looking back on your youth—with the newer generation generously mocking you. As for friends, I hope it is perfectly clear by now you are free to go against the flow and include personalities from the internet, rather than forcing yourself to deal with your newly divorced Aunt Liz or your teenage sister who just started menstruating this year. Yes, you've still got your best bud, Bill, who is still an avid nose picker in public at the age of 25! Who can compete with such a strong affinity?

Obviously, the perfect combo in someone's life includes not only internet personas, but also people from their own surroundings. Classmates and work associates, family and friends, people you go to the gym with, or who you fish with by the lake once every month—they all have the potential to form special relationships with you over time. The only requirement is that you keep an open mind and let them try. No matter if you have met them online or you have known them your whole life, you can try and see what happens. Just remember, you don't have to settle for that uncle who insists on telling the same joke over and over again in family gatherings. No one is laughing, Uncle Niels!

Business Is Taken to a Whole New Level

Businesses have been created with the sole purpose of generating money, right? Anything that escapes this norm should not be considered a viable corporate solution. Well, news flash! Open source software is what many well esteemed companies have been built on, offering a new opportunity to the business world. Instead of concealing the way they run, they maintain everything exposed online. In this way, whoever wants to check out code or gather information about a specific process, *no problemo*! What do you think about that? Do you believe somewhere in the world there is a hectic CEO that has lost sleep due to the anxiety of no longer keeping code a secret? Think again!

Businesses have grasped the concept that the future has nothing to do with isolated ventures or codes that mean nothing to those outside the company. On the contrary, once more the power is in the numbers. The more the people who work on a specialized task, the better. Another set of eyes is always refreshing, which is what I am coming to realize now that I am writing this book. Imagine writing code nonstop and missing just a tiny bit here and there. Disaster in disguise! To avoid that, the safest way would be to always work in pairs and have somebody run through the code to identify errors and typos. Imagine the both of you trying to fit in under the desk, coordinating your moves, and trying your best not to scratch your torso. OK, I see now that this would be impossible.

Now take a moment to figure out why open source software has taken the world by storm. You run a business and you want nothing more than be kept competitive, to stay one step ahead at all times. So why on Earth would you share your software with the rest of the digital universe? As it turns out, this would be a quite strategic move on your behalf. If you ask me, you should already have done so, assuming that your business can operate in the open source model. Adobe is doing it, as are Cloudera, Linux, and GitHub. Google and Facebook among others share their software and even help others fully comprehend what they are working on. But why? Why aren't they afraid?

The truth is there are many different benefits obtained by open source software. Indeed, they are so copious they leave no room for speculation or second guessing. It is needless to say that online you can gather the most aspiring minds of our time. Obviously, they cannot all work for you. And let's face it, who could afford them? However, with open source code you don't have to do that. With open source software you can still benefit from their brilliance and knowledge to evolve the code and progress your line of work. Whenever there is a glitch, a new challenge arises. Experts from different parts of the world study the code and suggest solutions, improvements, or tweaks that allow you to move forward.

Flexibility plays a key role to the expansion of your business and this is something you should not take for granted. Open source software can be modified as per your own customized needs. There are no restrictions, as you basically engage in the "trial and error" strategy. You can add code where you want more features or delete the ones that do not appeal to you or your business profile. Once it fits, you know you have done something right. Plus, if you find something doesn't quite cut it for your business in the long run, you can always change it. Investing all that time, effort, and of course money in premium software, on the other hand, would have been cost prohibitive.

Finally, I have left out one extremely important aspect of open source, which is what has made it so popular over the past few years among professionals, business entrepreneurs, and idealists alike. It is that sense of community that allows the corporate world to join together and contribute to something greater than individual needs, interests, and gains. For many people, achieving greatness means taking part in something that surpasses your selfish goals and setting the foundations for even more significant changes in the future. Who wouldn't want to help a fellow professional in a time of need? Who wouldn't want to feel that amazing sense of comradery, in a friendly and welcoming environment that stimulates growth and acceptance.

When professionals who are driven by their own passion and thirst for knowledge, growth, and improvement come together, the results are impressive. The open source community is out there, ready to offer a helping hand when you need one through a wide array of educational resources and documentation, tutorials and FAQs. Plus, whenever you feel like you are way out of your league, just dive into the chat rooms and forums with the pros. You will instantly feel at home and get things sorted out before you know it. How cool is that? Turning to colleagues, professionals of the same niche who may be miles away, and getting the answers that you have been searching for—easy, effective, and free.

The future of the corporate world has already been set in motion. Some years ago, people would have been reluctant to trust others online. It would be more sensible to take online information with a grain of salt. Apparently, you still need to do that. Nobody claims that humanity has been stripped of its lowest and vilest of instincts. There will always be those who feel it is their duty to scam you. But set that aside for a moment and visualize the big picture: a fellowship that puts common welfare before anything else and promotes sustainable success for everyone. How isn't that idyllic? Once you get past the personal interests and ensure that transparency is unconditional online, you will benefit from the cool features of open source wizardry.

So, forget what you have learned so far in life. Businesses are not restricted to a specific model that motivates them to compete with each other in a battle until the end. Sharing online is essential so everyone gets all the resources required for their own growth and, at the same time, contributes to the community. After all, it is not rare for experts to seek that level of recognition from their peers. They often pursue constructive criticism and assessment of their work. This leads to self growth and success for life, not only on a professional level, but also on a personal one.

The leading companies of the world have made that transition and have trusted open source with outstanding results and unique features. They have chosen to face the upcoming hardships and the challenges that are bound to rise over time together, supporting one another and allowing every business to take part in this fabulous collaboration. If you want to be part of the future, embrace this mentality in the way you run your business and you will soon feel liberated, ready to take on more challenges. Do not hold back and depend heavily on your limited means. The internet has taught us better than that!

VPN: Getting Rid of Space and Time Boundaries

People are in desperate need of a tool that not only allows them to protect themselves online, but also unlocks many wonderful features and bypasses all kinds of restrictions. Yes, there is a tool that does all that, along with some extra perks that you are about to discover right now. So brace yourself and get ready to learn a thing or two about the way the web works. If you haven't guessed it already (well, the heading should have given you a hint), we are going to focus on VPNs.

VPN is much more than a hype. Many people swear by it, as it opens up a whole new spectrum of opportunities to internet users all over the world. However, there are hundreds of misconceptions flooding the internet and this makes it more difficult for society to truly grasp its meaning and potential benefits. You will be surprised to see that most VPN software options out there are open source. This is yet again another sign that online sharing has rocked the digital world. But why is literally every internet privacy advocate suggesting we switch from conventional connections to VPN software? Why is it necessary to implement VPN whether we are traveling to a different country or simply trying to connect to a publicly shared Wi-Fi hotspot?

First things first. VPN stands for Virtual Private Network. VPN was established back in 1996, when PPTP was developed by Microsoft. PPTP stands for Peer To Peer Protocol. Does that sound familiar? This is a software that allows you to redirect your traffic and encrypt it through its servers. As a consequence, you do not have to jeopardize your personal information, log in credentials, or confidential documents online. The VPN does all the hard work for you and prevents hackers from snooping around in search of prey. Well, I surely hope this is a reason for you to reconsider using VPN, even if you are not that tech savvy. After all, there are VPN options that require little to no actual knowledge of the matter.

Thanks to the increasing internet speed now available in most parts of the world, people got the chance to know more about the benefits of VPN. Not many people enjoy having their sensitive information intercepted by the government or by marketers who want to know you better. There is an entire industry depending on specialized content for each internet user and you should resist them any way you can.

Personal information should be kept private and this is where the VPN excels. Going through the nodes and redirecting your traffic through the use of special servers, you don't have to worry about your IP address. You can get a randomly generated IP address in a number of minutes. In fact, you can fully customize it so that you appear to be located in Latvia or Morocco or

Italy. The only restriction is the locations of the servers in your VPN service provider's network. So prior to choosing which software you should download and install, take a moment and clarify these things. Apart from the instant boost of confidence you get from knowing that nobody knows your location, this option gets a whole new meaning in the digital world of entertainment. To be more specific, you can tweak your IP address accordingly and gain access to content that would otherwise be out of reach for you. Be it censorship from the authorities or from your own family, an internet provider or a service, you have a lot more possibilities now through the use of VPN.

A VPN is also able to bypass blockers that would otherwise slow your performance down. This happens because some ISPs reduce the speed of the internet once they spot you using it for downloading and sharing files, gaming, or live streaming. They do so, in order to tackle the challenge of having a network that is crowded with users (though you will never hear them say so). With the contribution of the VPN, you can forget about such restrictions and enjoy the maximum speed rate of your internet plan. More than that, here is an extra tip for maximizing your potential when using a Virtual Private Network. You can use your new IP address to navigate the websites you normally visit for booking flights and hotels. It might come as a shock to you, but the truth is you may get a generous discount.

By all means, protecting your computer and keeping those security breaches at bay seems fun. Well, maybe fun is not the best way to describe it. Nevertheless, through the use of a VPN you can enhance your overall online security level and not break a sweat. In fact, you can share files online for prolonged periods of time and maintain that same level of security and still have access to those files remotely. So even if you have forgotten a time sensitive document and you need to present it at an upcoming meeting, you can pop in and download it on the spot. VPN to the rescue!

With the help of the VPN, which is powered by open source software, time and space do not matter anymore online. You can be anywhere in the world and enjoy high-speed internet connection, and optimized security and performance at all times through state-of-the-art infrastructure. All that has been created by people who appreciate the importance of surfing the web in peace. It goes without question that the future is going to hold even more astounding services, like lightning fast internet and other amazing features. The VPN is certainly moving towards the right direction, lifting many restrictions that might have made your online experience less appealing and productive. If you have never tried VPN, I urge you to do so. But be careful not to drop your jaw, as you see what this technological wonder has to offer. Challenge accepted?

Chapter 4: What Keeps Open Source Going?

Technology has come a long way and it has transformed society to a great extent. Imagine a day without the internet, or without electricity, phones, apps, etc. Yes, it would be horrible! Yes, we would be able to appreciate nature more, but we cannot deny how heavily dependent on technology we are. Don't regard it as a bad thing, as it isn't. Ever since the beginning of time, man has been discovering new ways to improve his life. There is nothing wrong with fire, right? Where would we be now without its powerful effects?

Anyway, a lot has happened over the years and we have reached a point where technology is pretty much as vital to us as the air we breathe. The digital world has been responsible for some of the most amazing revelations in modern society. People from all over the globe join forces remotely and create true wonders of mechanics and technology which are bound to change the world as we know it. Sometimes these changes save lives. What could be better than that? Do not underestimate the power of technology, let alone the predominant feeling of selflessness that has driven people ever since antiquity.

You may believe this is all just theoretical wishful thinking, but the truth dictates otherwise. Through open source P2P file sharing, this dream becomes real and its chain reaction is absolutely marvelous. A tech guy from the Philippines collaborates with two Canadians and a Greek lab to create something that will rock the world of science. Does that sound impossible? It has most likely happened in the past or it is about to happen anytime soon. The web world is a majestic place, where boundaries vanish and people get together connected in a vast network. The old saying, “united we stand, divided we fall” gets a whole new meaning on the web!

Let’s have a look at some of the recent breakthroughs in technology triggered by open source file sharing and which have made a huge difference in our lives.

The Magic of 3D Printing

The Covid pandemic has been earth-shattering for the entire world. So where does open source file sharing come into this? Well, in a time of need the scientists did their part and rose to the challenge. With the global supply chain at a stand still, 3D printing was able to print face masks and shields on demand. Besides masks, 3D printing also produced replacement valves for resuscitation devices. In Northern Italy, a hospital ran out of these crucial parts and the fate of its patients relied solely on scientific initiative. With the help of innovative technology, doctors were able to overcome an enormous obstacle and manage the crisis more efficiently (Sher, 2020).

Obviously, 3D printing is essential for more than a pandemic! There are websites like Thingiverse that enable the sharing of digital design files. Millions of web designers exchange files and discuss innovations in 3D printing such as new printing materials and techniques. Such platforms encourage creativity and focus on optimizing the end product. Plus, there are educational resources to assist people in mastering the technology. There are also ongoing shared projects that promote collaboration. In case you were wondering about malicious intent, Thingiverse's founders are pretty straightforward and state that users should never upload or share content that "contributes to the creation of weapons, illegal materials, or is otherwise objectionable."

Bitcoin - Blockchain

Another prime example of P2P technology is blockchain, which is the infrastructure behind cryptocurrencies like Bitcoin. You all know Bitcoin, which was conceived in 2008 and was introduced to the public in 2009, implementing the open source software of blockchain. It has rocked the financial world, creating something nobody could have anticipated before then (can someone say Black Swan?). Cryptocurrency's popularity soon skyrocketed, establishing Bitcoin as one of the most acclaimed currencies ever in the world.

There are many benefits deriving from the use of cryptocurrencies and we have P2P to thank for that.

Bitcoin provides users with a very easy interface and allows them to minimize the cost of sending and receiving money. The banking system cannot support such a revelation and therefore the operational costs for wire transfers, etc. can be daunting. With the use of Bitcoin, though, you instantly send or receive money anywhere in the world. No authorization is required, which makes things so much simpler. Without middlemen, the whole process runs smoothly and is way more affordable for both parties.

Cryptocurrencies have taken the world by storm, thanks to their extraordinary perks. They were introduced to the world to satisfy people's growing need for low cost hyper-security. Globalization dictates people have access to their funds everywhere and any time, without paying a small fortune every time they want to withdraw cash. With the contribution of blockchain, money transfers are now easier than ever. The world has changed again and the wonderful world of P2P played a key role in this evolution.

There are who people claim there are drawbacks to cryptocurrencies that cannot be ignored. This is true. There have been some incidents regarding online theft over the years. But blaming cryptocurrencies is like condemning the bank because one of its branches got robbed in the past. The truth is any online system is going to be dependent on security measures. Apart from the ones taken by cryptocurrencies, it is also important users pay attention to online safety and not compromise with anything less than perfect coverage of their needs. In this way, the Bitcoin experience does not involve any serious risk.

The Innovation of Tesla

Elon Musk is difficult to describe. He is an engineer, philanthropist, visionary and industrial designer, net worth of over 83 billion USD in 2020. Having connected his name with SpaceX and Tesla, Elon Musk has been working towards revolutionizing transportation on Earth as well as in space. He is a complex and intriguing personality, whose every quote turns viral and offers food for thought for millions of fans all over the world. He has an impressive IQ and is definitely one of the most controversial entrepreneurs of all time. Tesla is aimed at creating electric vehicles that meet the needs of even the most discerning of drivers, whereas SpaceX is an aerospace company with an extraordinary vision.

Having all that in mind, let's take a look at the innovations Elon Musk has introduced in the global market: solar roofs that can convert solar energy into clean, pure energy for your home;, giant batteries for storing massive amounts of energy for commercial purposes; Powerwall, an emergency storage battery that provides homes with electricity during outages; and Tesla Models S, 3, X and Y, as well as the Roadster, the quickest production car in the world, with record-setting acceleration, range and performance. All these technological advancements are continuously updated, upgraded and optimized. So once you are lucky enough to acquire one of Tesla's creations, you are good for life.

How can you be sure you do not get left behind when it comes to upgrading the performance and features of your purchase? Elon Musk has thought about it and has reached a verdict. Back in 2014, he decided all his patents would be free for the industry to see. More than that, he announced he would not file a lawsuit against competitors using Tesla's technology. How cool is that? Despite the fact he was the one to come up with such exceptional masterpieces of technology, he decided this type of knowledge should be universal. It should be easily accessible to everyone who wanted to take advantage of the technology already in operation, so it could be more easily evolved into something even greater.

Apparently, Tesla's patents are free for the world to enjoy. Furthermore, open source hardware makes it possible for Tesla owners to constantly get the newest upgrades and benefit from even more amazing features. Technology runs wild, after all. The basic principle behind Elon Musk's decision is the fact huge discoveries are not a single man's property. If this were the case, then mankind would not have come so far. On the contrary, people have benefited from collective knowledge. As a result, large groups of heterogeneous individuals contribute their skills and acquired knowledge to evolve the groups even further. This has nothing to do with selfishness; individuals join their forces and offer what they can for the common good. From his point of view, Elon Musk wants to ensure scientists from all backgrounds can benefit from his discoveries with no legal prohibition standing in the way (Musk, 2014).

Yet again, some people might argue this knowledge can be used for malicious intent. This is why Tesla clearly states there will be no lawsuit against people who implement the specific technology in good faith. Otherwise, things will be quite different and Elon Musk will not give up his hard-earned rights. As long as a company using Tesla's patents complies with the Patent Pledge, they have nothing to worry about. Either way, sustainable transportation is on its way and Tesla has made remarkable breakthroughs in this field. By allowing others to make use of its patents and work freely to expand, improve, or modify them for optimal performance, Tesla presents a once-in-a-lifetime opportunity. Open source is crucial in reaching a much broader audience, collaborating with far more companies worldwide and updating the features of technological products.

High Speed Global Wi-Fi

Although technically this is another of Elon Musk's visions, I feel it deserves a special note. How many times have you found yourself in search of an internet signal on your phone? How many times have you wished your bandwidth was a lot more powerful so you could enjoy your internet connection to the fullest? The internet is not a luxury in today's world, it's a necessity. So it makes perfect sense why unobstructed, high quality Wi-Fi is on top of the list of technological priorities. By staying connected, you are able to experience the benefits of the digital world. Not only do you remain up-to-date with current affairs and news, but you also enhance your business and optimize communication, entertainment, and information gathering about literally everything across the globe.

However, even today there are places where you get no Wi-Fi signal whatsoever. You seem to be forgotten in time, unable to claim what is yours online. Areas that are hard to reach, mountainous terrains, regions suffering from extreme weather conditions, remote islands and villages without the proper infrastructure, are just a few of the places where the internet does not thrive. It is the equivalent of not having electric power for your business, while the country still expects you to pay your fair share of taxes and compete on the same ground. This simply cannot be. And of course, Elon Musk has come to the rescue!

SpaceX has been forming a satellite constellation in order to overcome the boundaries and create the perfect infrastructure for high speed Wi-Fi internet. With the help of satellites, Starlink will provide the world with unbelievable Wi-Fi speed rates. At the same time, the innovative company has focused on making the satellites as compact as possible and keeping space clean. But the internet coverage will not be limited to the areas where it is hard to get a decent signal. Instead, there will be internet plans that meet the increasing demand for internet access in cities and towns. too. Elon Musk's plans are not even restricted to planet Earth. His long term goal is to serve the internet needs of Mars, which is why he has been working wholeheartedly to achieve lightning fast internet speed rates all over the world.

Understanding the web is something people cannot do without first providing excellent internet service. Obviously, there are many barriers that need to be overcome until everyone gets a quality Wi-Fi connection. Still, it is great there are plans aimed at developing global Wi-Fi. This is definitely a concept that overlaps with open source, isn't it?

Chapter 5: In Anticipation of What Lies Ahead

The future is filled with exciting things! Every minute of the day, technology changes and offers us wonderful opportunities to witness its grandeur. The future is racing towards us much faster than we would have expected. Are you a fan of sci-fi? Then you are already counting down the days until the moment you get to experience the future. In fact, every year brings you closer to what you have been dreaming of.

From Sandra Bullock and *The Net* (1998) to Keanu Reeves and *The Matrix* (1999), from Sylvester Stallone and *The Demolition Man* (1993) to Jeff Bridges and *TRON* (1982), or even James Woods and *Videodrome* (1983), science fiction is always meant to excite the senses and allow you to dream of a future where all that technology is real. The time is now to start experiencing what you have been visualizing for so long in your mind.

Of course, there are things that have been left solely to our imagination. I don't think Bumblebee the Autobot is ever going to switch radio stations for me, although I haven't given up quite yet. Optimus Prime is a great guy and Megatron is the epitome of villains, but this is all just fiction. Still, technological advancements seem to be closer than ever. Nowadays, automated systems are not limited solely to the elite. Instead, everyone is able to save up money and install a system that allows them to control their home locks, adjust heating and cooling, and monitor their rooms and backyard.

It all boils down to innovation and covering the contemporary needs of society. In a world where nothing should be taken for granted, it makes perfect sense for scientists to go the extra mile and envision what the world is going to look like in the future. How is the structure of society going to be different and what changes will that set into motion? Technology must be one step ahead and dictate trends, anticipating developments instead of adjusting to them accordingly.

Back in the day, when someone acted like a visionary, he faced the threat of doubt and rejection. More than that, he faced the threat of excommunication or even execution. People used to regard those who differed from the norms as dangerous and they identified them with the very essence of evil. How can somebody envision what is not there? It doesn't make sense. In their minds, people should comply with the political establishment and not stray from the status quo. Otherwise, they posed a danger to society and needed to be dealt with as soon as possible.

Fortunately, things have changed a lot since those dark times. Today, a visionary is embraced and praised for thinking outside the box. There are no norms to abide by, let alone rules to follow. The sky's the limit and this is why so many great technological advancements have taken place. Nothing can stop us now, right? Even though it is always good to be humble, it is even better never to cease pushing forward and searching for more. Besides, technology is bound to flip several industries upside down and I can't wait to see what's coming next.

Can you?

Safeguarding What Actually Matters

Safety is of utmost importance to us all. We all go above and beyond to protect what we hold dear in our hearts. We build our life around a house we call home, making sure it keeps us protected and safe from harm. This is our shelter, the place we turn to even if everything else falls to pieces. In the same manner, we make sure all the places we spend time in are equally safeguarded. Be it a company headquarters or a cottage house, a shopping mall or a school, we pay great attention to creating solid buildings and implementing state-of-the-art systems that improve their overall safety. Technology has given us some truly powerful weapons, so we can strengthen our security and maintain our safety at all times. And the future shines even brighter, showing the way for even more elaborate systems to be put into effect.

We have upped security in several aspects of life. We already know we can use alarm systems and automation to our advantage. There are special gadgets that enable us to monitor motion, as well as temperature and humidity. Furthermore, cameras are connected to smartphones and most mobile devices, facilitating our remote control. Have you forgotten to lock the back door? No problem. With the click of a button, you can lock it from your office. Do you want to see if you have left the window open? Check the locks and connect to the specific room in question. Do you want to let your dog out in the yard? Unlock the kitchen door and do so without jeopardizing the safety of the entire house.

The future will hold many surprises, which will aim at safeguarding society even under the toughest conditions. Technological advancements have produced exceptional machines, systems, and networks that work in perfect alignment for the optimal performance. Artificial intelligence (AI) is everywhere. Peter H. Diamandis, founder of XPRIZE, wrote: "Another example sits at the convergence of AI, networks, sensors, and satellites. Here, we gain the ability to develop global threat detection networks far more sophisticated than anything that exists today. Suggestions run the gamut, such as global food web monitoring to protect against catastrophic famines or terrorist attack; atmospheric sniffers that hunt for everything from plague-causing pathogens to the scent of nuclear materials; and rogue-AI detectors—essentially AI built to hunt for rogue AI." (Diamandis & Kotler, 2020)

Rogue AI?! What? AI that hunts AI?! Wow. Talk about sci-fi becoming reality. I mean, watch Terminator much? Preparing for the unknown is crucial, since there will always be threats we can't even imagine. Even if you tried to foresee what might happen based on evidence and trends, you still cannot be 100% certain. Life gets in the way when you least expect it and turns everything upside down. Sound familiar? How can we protect ourselves? While no one will ever be certain what threats the future holds,we do know that open source future technologies will give us the greatest chances of survival.

If you are better prepared and if you have adequate supplies, infrastructure, and facilities to cope with the abrupt changes that will take place, then you can survive. Otherwise, imagine trying to breathe in a toxic atmosphere without having the necessary breathing apparatus. You would be doomed to die right from the very first minute. But if you have the equipment required to breathe and walk, without any fear of toxic burns on your skin, etc., then your chances for survival are much greater. Alternatively, think for a moment you fall down from the sky, with your eyes fixated on the ground. Adrenaline has spiked, but you realize you wear nothing but your casual outfit for work, jeans and a T-shirt. This is shocking. Now think you check again and you notice a parachute. You are not helpless. On the contrary, it is your resourcefulness that has given you the opportunity to survive—or at least to try!

Hope for the Future

The French economist Thomas Piketty believes not all people are given the same opportunities to access information. He points out that current banking regulations do not allow for proper taxation of the ultra-wealthy because much of their income information is distorted or completely hidden. (Hussey, 2014). This leads to inequality, since there is no spherical knowledge as to resources. Piketty goes on to state the obvious. “The principle is quite simple: national tax authorities should receive all the information they need to calculate the net wealth of every citizen.” Can you imagine that? Every citizen has a net worth and the authorities should make sure this is accurately calculated.

“There's nothing in the world so demoralizing as money,” said the Greek tragic poet Sophocles. The majority of men will be corrupted, once they get their hands on the right amount of money. This is disheartening, to say the least. Something as plain and simple as money can make the world go round, affect consciousnesses, and determine who gets to live well and who merely survives. For many centuries, in fact since the beginning of time (with a hint of exaggeration), money has become a matter of controversy for societies. In several societies, people who had the means and the resources would go up the ladder and become powerful. Then, that wealth and power became hereditary and the families become aristocrats. It took a lot of hard work, almost unbearable sacrifices and struggles to turn things around.

Money creates influence and increased social status, as I am sure you are aware. But who owns the money? Even though you think you’ve got the answer to this question, reality will come as a shock. Most of us do not have a clue! Access to such information is crucial, as it allows the world to have a clear image of society’s financial status. Are there thriving citizens or is there intense financial struggle in specific parts of the world? Many different groups of people suffer from discrimination when it comes to pay. This should be pointed out so the authorities can do something about it and reverse the current situation. Without knowing where the problem lies, though, there is not much to do. At best you can speculate and conclude as to what causes the problems in the economy.

It goes without saying money should always be kept in the light. You are not trying to divide anyone, let alone make matters worse by hiding the inequality of wealth. There is no point in escalating the differences among groups of citizens based on their income. More than that, it should be perfectly clear money is not to be demonized. However, it needs to be declared in full and reflect the economic status of each citizen. Otherwise, society cannot know right from wrong. How can you expect the world to move forward, unless reality is depicted in its true colors?

Over the past few years, financial scandals have shocked the world by revealing the depth of corruption. Yet, this information has a silver lining. It allows everyone to understand the inequality and conflicting interests that get in the way of creating a better society. Unfortunately, there are too many people who take advantage of their rank or power for their own gains. They do so with the tolerance of others. Still, if the culprits were brought to justice (or at least held accountable in the eyes of the public), things can change.

The Panama Papers were an epiphany, shedding light on innumerable shady financial transactions.They revealed international schemes, offshore companies, tax evasions and many many other problems in the banking industries.

Information can change the world. It can begin the change that will root out financial corruption, correct wealth distribution, and ultimately eliminate poverty worldwide. Open source sharing is key to this ongoing effort. Having witnessed just how badly rotten the system is, there is no way citizens will remain silent. Now it is the time to act, with unbiased transparency and raw truth. The web has been the means by which information has traveled quickly and efficiently to all parts of the world. Peer to peer, end to end, distribution of files has set the foundations for something great, enhancing the hope for the future.

A Struggle for Internet Neutrality and Anonymity

One of the best things about the web is you can be whoever you choose to be. Again, don't get me wrong. I am not a fan of identity theft, nor do I support those who rant and rave simply because no one is able to hold them accountable. It is true that lack of consequences can be a mixed blessing for society. Yet, the internet is a great place for those who value their privacy and want to remain anonymous. Not everyone dreams big, nor are all people in the mood to constantly look over their shoulder when surfing online. But sometimes you just need to take that identity off and feel the weight lift off your shoulders!

Surveillance, whether by the government or by third parties, can make you feel extreme discomfort. It is not reassuring to realize someone is tracking your digital footprint, analyzing your every move, and judging you based on your online preferences. I don't want to be the one to break it to you, but this is what happens pretty much all the time on the internet. Someone is watching you unless you make sure you use end-to-end encryption to cover your tracks properly. There is a whole industry depending on your digital self, and internet privacy just leaves them hanging. Are you thinking about messing with their plans?

Some people have had enough with the invasion of their privacy and their right to remain anonymous. Steve Huffman and Alexis Ohanian, who founded Reddit back in 2005, are firm believers in the right to privacy, and they joined forces with many other like-minded internet entrepreneurs. They have been active in the field over the years, fighting for net neutrality rules and against any discrimination preventing proper use of the internet. Net neutrality dictates internet service providers (ISPs) treat all users in the same way, regardless of the content, the destination website, or any other element that might affect objectivity. In this way, users do not have to stand by while their speed slows significantly while torrent file sharing or live streaming.

Huffman's and Ohanian's activities have had a substantial impact on the digital world. In 2012, Ohanian also co-founded the Internet Defense League to defend net neutrality. If you want to evaluate their effectiveness as a portal of unbiased information, discussion, and rating, check out the countries where Reddit has been permanently blocked. And in case you are a fan of statistics, you will be happy to know that in 2020 there were about 430 million monthly active Reddit users, with over 1.2 million different subreddits. Those numbers are pretty darn amazing.

Looping Back to the Beginning

Isn't the web an amazing place? Filled with surprises and excitement everywhere you look. There are thousands of different things to do, people to meet, hobbies to take up, professions to follow. You can chill or search like crazy for information, hang out with friends or enjoy some wonderful music online. One of the fundamental purposes of using the internet in the first place was communication. Take a moment and think about the Napster heroes, Shawn Fanning and Sean Parker. They were both tech-savvy youths, immersed in a majestic world and looking for something that would change their lives. What happened? They met online, using Internet Relay Chat (IRC)!

An underground internet chat room is what triggered the entire revolution of the music industry. IRC is an example of similar people sharing information and insights. There are many different options out there, depending on your likes and preferences. Reddit is one of them, even though this discussion only covers part of its use. So what is your poison? No judgment here! There are forums literally covering all types of interests, from fishermen to amateur photographers, to those who want to decompress after a long workweek in a VR chat room while sipping a drink at home. Obviously, IRC's glory days have passed. However, this doesn't mean there are not many fans who are still hooked on this nostalgic interface.

In the past, online communication has taken various forms. Back in the day it was common for internet users to use BBSs, or Bulletin Board Systems. Let's face it, you would probably prefer to watch paint dry than use a BBS on a daily basis. But it was cool and it was perfect for dial-up connections to the internet. Chat rooms, of course, have had their fair share of success over the years. Imagine a super slow Facebook Messenger without all the emojis and there you have it. Oh how far we've come!

And who can forget about the humble email? Having been conceived as a means of online communication ever since the 1960s, email gained unbelievable popularity. How could it not? It is simple, easy, and fun to mail something and have it delivered on the spot! No more stamps, no more long queues, no more lost mail that cannot be retrieved. It was a dream come true for everyone. Its applications in modern life are endless. You can use it to subscribe to a newsletter or create a social media account, send and receive corporate files, promote your business, share cool stuff with friends and family, or simply gather all the latest ads in any given industry. Email accounts are free and everyone has one—at least!

Back to Napster and its founders. How can an interaction like Fanning's and Parker's lead to a company that changed an industry forever? Which planets aligned with one another, so those two teenagers started communicating via IRC? Realizing a single decision you make can affect the rest of your life is truly fascinating. If it hadn't been for that primitive chat room, Napster might not have been born. And as a result, maybe the music industry would not have crumbled and fallen due to the massive loss of income. Free file sharing with a specific focus on music files might have been conceived, but when? The timeline is not clear.

If Napster's founders met entirely by chance, what does luck hold for you in the future? Who could *future-you* meet in a virtual reality chat room of tomorrow? It makes sense that communication will only get better and better over time. Technology has created some true masterpieces of design, able to bring us closer and awaken the senses in ways we would not have thought possible. So which enticing encounters are waiting just around the corner for you? Are you going to change life as we know it simply by going online and talking to someone who gets you? In the past, this would be the equivalent of running into Socrates or Da Vinci and starting to talk about things you both like. What kind of flame would that discussion ever spark?

The advancements of the future are nearly mind blowing. Distances are no more than numbers now; physical barriers are meant to overcome without delay. There is nothing holding us back from connecting with people half a globe away, chatting like we are sitting face-to-face and sipping on a mug of warm green tea. Communication is effortless, words flow smoothly, and you narrow your audience per your likes. Nobody is telling you who to talk with or who to avoid. Unless there is a restraining order issued by a court, you are good to go! Let that chatty fellow inside you bubble up and make some new friends. And who knows? It may be written in the stars your startup co-founder is waiting for you to go online. You wouldn't want to keep him waiting now, would you?

Conclusion

Are you familiar with the Manhattan Project? During World War II, there was an overwhelming fear Nazi Germany would use nuclear weapons to prevail. In response to that threat, the U.S. organized the Manhattan Project to bring together the most outstanding minds in the free scientific world to harness the energy of the atom before the Nazis did. Although we are talking about a controversial invention that continues to pose one of the greatest threats for the world, we should focus on the collaboration that was developed back then. In light of a new common threat, people had no other solution but to stand united and try to deal with what was coming. There was no room for selfishness and personal agendas, and there was no time to waste. Scientists worked day and night, building the weapon that would provide them with an extraordinary advantage.

The Manhattan Project was not only innovative in terms of collaboration, but also extremely secret. Very few people knew all the details about the project, and all participants were sworn to secrecy. According to Life magazine, "probably no more than a few dozen men in the entire country knew the full meaning of the Manhattan Project, and perhaps only a thousand others even were aware that work on atoms was involved." This shows just how well the cogwheels turn in a perfectly oiled machine aimed at saving the free world. By all means, mistakes were made along the way. Nobody claims the Manhattan Project was flawless. In fact, some people claim this was the driving force that pushed us all into the Atomic Age and its irreversible repercussions. This was one of the first times the Americans and the British worked together. What united them was the mutual threat posed by Hitler and the Nazis. Otherwise they would have persisted in identifying their differences and not spotting their similarities. This is the exact opposite of a productive mindset. As soon as the world coordinated their actions towards achieving their goal, no barrier was high enough to keep them away. It is no wonder why true power often comes in numbers. Unless you find your allies and set your priorities straight, you can never expect to win in the long run.

Being able to overcome differences and cooperate with others due to shared vision for changing the world is an example of social maturity. Society is not driven by its impulses and instincts. It does not behave irrationally, motivated by short-term profits. Society has grown and is now capable of understanding what actually matters. No man is an island, nor should we be regarded as such. Humankind was not meant for isolation and we all have a lot to gain from collaborating and living together in harmony. Besides, take a look at most of the things that have been used to divide us. Are they really worth it?

True AI and the open source collective consciousness is upon us. Conditions are now ideal. Internet users are savvier than ever and they have already become part of that wonderful gift. Information is shared online without bias and is easily accessible by everyone. Given that, it makes sense why society is ready to move forward and renounce past habits that have held us back. It is time to set aside trivial things that appear to be rising up like mountains. The danger lying ahead of us is to allow ourselves to be fooled into thinking whatever obstacles we encounter cannot be overcome. That is not the truth. In a mature society where the concepts of freedom, equality, and openness prevail, things are bound to improve. People will discover how good it is to share. They will join groups that are based on love and intelligence. These groups will be like new families, established within the society to protect us from all harm. Groups created without conditions will join the battle against hate, terrorism, and chaos. And they will most likely win in the end! A just cause motivates people to try even harder, until they reach their end goal.

We want to form groups with those of similar interests. Every mind whose philosophy is close to ours is welcome to join. The more, the merrier! This is going to be the beginning of something great, something that will probably trigger a chain of reactions. And I am talking about the cool type of reactions, not the shady ones. The "global group" we need to gather is going to share information to honor human progress and the advancement of a global tribe that lives in a macro-version of homeostasis. Do you think this is just too good to be true? I am here to convince you otherwise. Hopefully, sooner than later you will see it for yourself and you will comprehend this mentality in its entirety.

If you wonder how open source P2P can ignite a ton of

fireworks at once, then you seriously underestimate its value. This is an act of selflessness driving the world further up the steepest road, and making sure it holds. It is an act of kindness that allows others to experience happiness and satisfaction, even though you may not necessarily feel that way. P2P is truly a gift, proving to the universe that the human race needs connection and sharing.

The Verdict Is Out: Sharing Is Caring

Can we live in a society where everything is transparent and free? The internet is testing our limits in that. Freedom of thought, online distribution of ideas and zero tolerance against corruption, these elements have been proven fundamental in modern society. Through the web, citizens become more actively involved in politics and develop critical thinking. They gather information from unbiased sources, they cross-check data and they express their opinion freely. Physical boundaries are disappearing and information is sent and received at amazing speed rates. We are living in a more open world, filled with potential beyond our wildest dreams.

Sharing is caring, which is embodied in the way people form groups, forums, websites, and blogs to communicate freely. There you will see unconditional help and sincere gratitude among users. It doesn't matter if they earn something out of assisting a fellow user. They do it anyway, because they feel good about it. Plus, they know that, sooner or later, it is going to be the other way around. They promote equality, through their own behavior. It is not just words; they make it happen.

Online information belongs to the people and should be freely accessible to everyone. Open source dictates just that. You can access the content and evaluate it, test it, and modify it accordingly. In this way, everyone becomes responsible for the acquired knowledge as it continues its journey across the web. Businesses thrive on such models, which allow them to maximize the talent of their scientists while maintaining minimal operational costs. The same goes for apps and tools that are free to download on the internet, and are always upgraded to fix glitches, viruses, and other anomalies. The internet is a living, breathing organism, and therefore it is constantly evolving and transforming. The only way to keep up is to maintain connectivity with the web and allow updates in the software you use.

Solidarity is one of the best aspects in modern society, and the internet is eloquent proof of that. People go above and beyond to ensure there is an abundance of resources on the web for everyone to benefit from. Even if they have covered their individual needs, they still work towards guaranteeing resources for people whom they have never met. This is the wonder of the internet, people connecting and collaborating in harmony. There are real treasures found online, shared in distinct formats and featuring a wide variety of files. From music to books, from paintings to 3D designs, from films and series to documentaries and recorded concerts.

A Digital Bill of Rights

Would you believe we are under attack by marketers? If you have heard about the 2018 Facebook-Cambridge Analytica scandal, this is not news to you. Facebook allowed the British consulting firm to harvest immense quantities of data about Facebook users. Apparently, no one has called me to ask for my approval on that and I am guessing you have not received that phone call either. So Cambridge Analytica used that information acquired from Facebook to conduct psychological warfare. Yes, ladies and gentlemen, we *are* under attack!

Through a sophisticated system of algorithms, the company had the chance to analyze the data and categorize Facebook users according to their preferences, likes and beliefs. So this means people who had a specific affinity towards pets, per se, would be categorized in a group and others who didn't quite fancy them would go in a different group. This is just an example, of course. The variables were myriad. In the 2016 U.S. presidential elections between Hillary Clinton and Donald Trump, such information came in handy. People were manipulated, being shown different types of content, videos and advertisements aimed at manipulating their voting decisions. The most shocking thing of all is evidence such manipulation influenced the result of the election.

This psychographic targeting of ads is real; there are personality profiles that each individual fits in. However, secretly working to influence the views and votes of over 50 million U.S. citizens can only be described as creepy and malevolent. Propaganda can be spread online and the public might not even realize it. Online information has definitely made it easier to unmask the scandal and address the issue. Still, do you think this is the first time consulting firms used such psychological manipulation? How many politicians have hired these propaganda pros so they can promote their own agenda and smear the reputation of their opponents?

Team Trump never really admitted to manipulating the people of the U.S. How could they? Instead, they insist they simply acquired the data from the consulting firm and then used it for their campaign needs. It doesn't make sense, if you ask me. When you have true experts in that field and you are already doing business together, why go elsewhere? Nobody can question how experienced Cambridge Analytica was at the time. Just take into consideration the fact they have been involved in well over 200 elections across the globe, including Argentina, Trinidad and Tobago, Nigeria, Ukraine, and the list goes on. So they obviously presented their new clientele with an impressive portfolio of results.

A former Cambridge Analytica Brittany Kaiser described in detail the shady tactics used by the company. According to Kaiser, Cambridge often used a "sophisticated infrastructure of shell companies that were designed to funnel dark money into politics" (Cadwalladr, 2020). The scandal forced Mark Zuckerberg, founder and CEO of Facebook, to apologize in front of the British Parliament, and the people of the U.S. and the world. He claimed there was no malicious intent on behalf of Facebook behind the collection of third party data, but promised to introduce full transparency from on the way such collection is performed.

Data driven microtargeting techniques might sound strange to you, but in reality they are here to stay. Cambridge Analytica got involved in the political scenery of developing countries so they could test their data collection products. If they performed well and the results were to their satisfaction, then they could move forward and implement the very same products or variations in other countries. In other words, they wanted to "test drive" the features of their services without any respect to the countries whose fates they messed with.

A Whole World of Data

As you see, data centric companies like Facebook and Google have been using your data and you may not even know this. It might seem harmless to share your opinion on a product or a service, leave feedback, and review a book or a film. Moreover, it may seem trivial to accept cookies that are meant to optimize your user experience online. This is a bait most websites use in order to make users feel comfortable. After all, tracking bits and pieces of data used on the web is not *supposed* to help website owners. No, everything is done to make you experience the internet to its maximum capacity.

How volatile is your digital life? How exposed are you against such acts of manipulation and data collection without your knowledge? I bet you are starting to doubt your every online move. Well, take a deep breath and relax. Things are not always black and white. There are many other colors in between offering that lovely rainbow, the sunrise and sunset, flowers, and all the other masterpieces of nature. So should you stay idle and watch them as they take advantage of your information to make profit? *No way.* Instead, you should take action and make sure this does not happen to you.

Things would be very different if these data centric companies decided to let us in and share their intent with us. For instance, when you scroll down your newsfeed and come up with cute babies eating and laughing and messing around with puppies, a pop up window could appear to show you the company is collecting data from you. More than that, they could offer you a buck or two to allow them to access such information about your likes. In that case, the decision would be yours, not theirs. You could let them have that piece of info about you or not, but either way you would be aware they collected it.

Let's take Apple, for example. The company has a colossal database of fingerprints and retina scans, right? This database must be featuring millions of internet users worldwide. As advertised, iPhones have been introducing facial recognition and two factor authentication as part of their attempt to enhance security. This is all great stuff, no doubt about that. But really, shouldn't we know more about what Apple does with these biomarkers? Maybe they are used to collect data for the authorities or even for some kind of experiments. I know, conspiracy theories are not that solid. More often than not, you need to get a grip and accept it is just what meets the eye. However, in a turbulent society where scandals pop up like mushrooms and multiply by the thousands, what makes you reject that theory and rest assured of your anonymity online? (Srinivasan, 2020)

It is disturbing to realize you have been cast aside due to one of your personality traits or due to the color of your skin, your marital status, or your geolocation. This is not how the world is supposed to be. Inequality does not belong in the digital world. The internet should be a safe place where everyone has free access and can benefit from a wealth of information. It goes without even saying this is nothing short of a dreamland for most people. Being able to rely on the web for impartial information and news updates that are not manipulated by those in power sounds amazing. If it is topped with equal treatment from the marketers and the companies that appear to be monetizing our every breath, then so be it.

On the other hand, who doesn't love convenience? Who doesn't want to whip up an online order in just a matter of seconds, thanks to Amazon or Google? These things make our lives easier, but at what cost? Spending a few more minutes while ordering or searching for similar products to the ones you usually buy on the web might seem boring. It most certainly is. But come on, who has ever jumped for joy because you had to buy diapers or detergent? It is part of your routine, so you can live with that. If you had to sacrifice either your comfort or your anonymity, which one would you choose? Either way, transparency is a must and the web cannot do without it.

New legislation needs to be put into effect reflecting current changes on the internet and addressing the hot topics which still bring controversy to the table. A digital bill of rights is essential, since this is the only way for everyone to safeguard their online experience. In this way, changes in bandwidth speed will not be tolerated, and ISPs will not have the right to interfere with your internet connection with the purpose of discriminating among internet users. We are standing on the brink of radical changes on the web. Now is the time to unite in order to claim what we are entitled to. If we stay passive and settle for whatever others decide about us, we really don't have any excuse. Inclusivity and democracy calls for us to take our theories and put them into practice!

In a Nutshell

You finally made it to the end of this book. I can't thank you enough for trusting me. I am sure by now you have come to appreciate the true value of online sharing. Open source is a means to promote collective consciousness and bring the internet to the world without bias or secret agenda. The web is a place for everyone to access. A place that allows mankind to evolve further and a harbor that promotes collaboration in a healthy and productive environment. It is time to join forces and form social relations that will change the way we think of businesses, money, equality, and happiness. With high quality Wi-Fi for everyone being just around the corner and with the majority of the land and sea already "internet ready," there is no moment to waste.

Throughout these chapters, you have read through some of the most exciting news about open source. The vision of a few *can* change society. Even though you may not notice, open source file sharing is everywhere and affects our life in multiple ways. Think of 3D printing and the majestic world of sharing designs, evolving the trade and even contributing to society's needs selflessly. Then think of digital wallets and the opportunity to transfer money all over the world in a blink of an eye. How many inequalities can be prevented through this simple innovation? And of course, the ingenuity of some of the world's leading influencers, such as Elon Musk, is more eloquent proof of open source scalability. The future is now and Tesla is helping pave the way. Will you stay behind or keep up?

Ever since Napster was conceived by two teenagers discussing in an IRC chat room, a lot has changed. The entire music industry has been shattered by the whooping sounds of P2P file sharing. Music was made accessible to everyone and the chain reaction has been mind blowing. Although the company was short lived, nobody can argue with its impact on the digital world. And let me tell you one thing about Napster. It paved the way for absolute online freedom and a shift in the collective mindset of people.

No, the online experience is not always going to be a breeze. How could it? There are dangers out there for you to watch for. There are people who will go to extremes to you or stall your progress. You need to accept that. The world is not all flowers and sunshine. No, at times it can be a battlefield. And if you want to survive, you need to adapt and be ready to tackle the threats. This is why it is critical (and I cannot stress that enough) to educate yourself about online security and what it actually means. I want you to think of this book as a general outline of the game. It's like American football; I've given you a guide about the rules and latest protective equipment. It's now your choice on what position you want to play and what your style is going to be. Everyone's included.

But all these struggles and all these challenges are only here to make us stronger. They are here to show us the future of the web is based on open source. Forget what you have been told, and embrace information sharing online with this newly given philosophy. It is liberating to feel the digital world knows no boundaries. The power of the people can actually flip the world upside down. Businesses do not still build their rep on introversion; they collaborate with the brightest minds in the field and together they push the industry forward. It is an exciting feeling, to know you can rely on your peers and expect quality results without prejudice. This is the material dreams are made of, and open source is the centerpiece.

So my wish would be to keep your horizons broadened and expect the unexpected, always believing in the good that lies within us all. The internet has taught us it is feasible to obtain that precious diversity and maintain its magnitude, while still fortifying our connective bonds. Possessions would mean nothing without someone to share them with. Am I wrong?

My next goal is to help provide you with more technological insights and philosophies of our modern day. Let's build a community of open source enthusiasts and see what we can accomplish!

Send me an email at nielsstarlingbooks@gmail.com to connect, or give me some constructive feedback about the book. Or heck, even if you just want to say hi

Thanks for reading and keep collaborating!

- Niels

References

Peer-to-peer piracy on university campuses. (2003, February 26). .Govinfo.gov. https://www.govinfo.gov/content/pkg/CHRG-108hhrg85286/html/CHRG-108hhrg85286.htm

#1 Password Manager & Vault App, Enterprise SSO & MFA | LastPass. (2019). Lastpass.Com. https://www.lastpass.com/

1Password. (2013). 1Password. https://1password.com/

Barnes, T. (2015, June 1). 16 Years ago today, napster changed music as we knew it. Mic. https://www.mic.com/articles/119734/16-years-ago-today-napster-changed-music-as-we-knew-it#:~:text=Napster%20forced%20the%20music%20industry

Braga, B., Druyan, A., Pope, B., & Dart, K. (Directors). (2014, March 9). Cosmos: A spacetime odyssey. Cosmos Studios.

Cadwalladr, C. (2020, January 4). Fresh Cambridge Analytica leak 'shows global manipulation is out of control.' The Guardian; The Guardian. https://www.theguardian.com/uk-news/2020/jan/04/cambridge-analytica-data-leak-global-election-manipulation

Cimpanu, C. (2020, January 22). Microsoft discloses security breach of customer support database. ZDNet.

https://www.zdnet.com/article/microsoft-discloses-security-breach-of-customer-support-database/

Diamandis, P., & Kotler, S. (2020, January 31). Here's what might save us from cyberterrorists, biohackers, and sentient robots. Linkedin.com. https://www.linkedin.com/pulse/heres-what-might-save-us-from-cyberterrorists-robots-diamandis/

Ferriss, T. (2015, December 28). Derek Sivers Reloaded – On success habits and billionaires with perfect abs (#128). The Blog of Author Tim Ferriss. https://tim.blog/2015/12/28/derek-sivers-reloaded-on-success-habits-and-billionaires-with-perfect-abs/

Gordon, S. M. (2005). *Downloading copyrighted stuff from the internet: Stealing or fair use?* In Google Books. Enslow Publishers. https://books.google.com/books/about/Downloading_Copyrighted_Stuff_from_the_I.html?id=GY482a64U5kC

Green, J. (2019, December 22). Chuck Palahniuk | Adjustment Day (W.W. Norton & Company). http://Theartsstl.Com/. http://theartsstl.com/chuck-palahniuk-adjustment-day-w-w-norton-company/

Gressin, S. (2018, December 4). The Marriott data breach. Consumer Information. https://www.consumer.ftc.gov/blog/2018/12/marriott-data-breach

Hussey, A. (2014, April 13). Occupy was right: capitalism has failed the world. The Guardian; The Guardian. https://www.theguardian.com/books/2014/apr/13/occupy-right-capitalism-failed-world-french-economist-thomas-piketty

movieweb. (2012). *Downloaded - official napster documentary trailer* [HD] Shawn Fanning. In YouTube. https://www.youtube.com/watch?v=6Ai6K2VIEXM

Musk, E. (2014, August 19). All our patent are belong to you (sic). Tesla.com. https://www.tesla.com/blog/all-our-patent-are-belong-you

Pew Internet Project. (2000, June 8). Pew internet project survey shows 13 million music "freeloaders" on the internet; 1 billion free music files among Napster users. Pew Research Center: Internet, Science & Tech. https://www.pewresearch.org/internet/2000/06/08/pew-internet-project-survey-shows-13-million-music-freeloaders-on-the-internet-1-billion-free-music-files-among-napster-users/

Pina, A. (Director). (2017). *Money Heist* [Streamed]. Atresmedia & Vancouver Media.

Rose-Stockwell, T. (2017, July 28). This is how your fear and outrage are being sold for profit. Quartz. https://qz.com/1039910/how-facebooks-news-fe

ed-algorithm-sells-our-fear-and-outrage-for-profit/

Sher, D. (2020, March 14). Italian hospital saves Covid-19 patients lives by 3D printing valves for reanimation devices. 3D Printing Media Network. https://www.3dprintingmedia.network/covid-19-3d-printed-valve-for-reanimation-device/

Software Updates. (2019, March 28). Tesla.com. https://www.tesla.com/support/software-updates

Srinivasan, R. (2020, January 28). Americans need a "digital bill of rights." Here's why | Ramesh Srinivasan. The Guardian. https://www.theguardian.com/commentisfree/2020/jan/28/americans-need-a-digital-bill-of-rights-heres-why

Voss, C., & Raz, T. (2016). *Never split the difference: Negotiating as if your life depended on it.* In Google Books. Random House. https://books.google.com/books/about/Never_Split_the_Difference.html?id=RmdqCgAAQBAJ

Wikipedia Contributors. (2019a, February 16). Napster. Wikipedia; Wikimedia Foundation. https://en.wikipedia.org/wiki/Napster

Wikipedia Contributors. (2019b, March 14). WikiLeaks. Wikipedia; Wikimedia Foundation. https://en.wikipedia.org/wiki/WikiLeaks

Wikipedia Contributors. (2019c, March 16). Edward Snowden. Wikipedia; Wikimedia Foundation. https://en.wikipedia.org/wiki/Edward_Snowden

Wikipedia Contributors. (2019d, April 27). Self-fulfilling prophecy. Wikipedia; Wikimedia Foundation. https://en.wikipedia.org/wiki/Self-fulfilling_prophecy

Winder, D. (2019, September 5). Unsecured facebook databases leak data of 419 million users. Forbes. https://www.forbes.com/sites/daveywinder/2019/09/05/facebook-security-snafu-exposes-419-million-user-phone-numbers/#59ae9ea71ab7

Made in the USA
Coppell, TX
04 November 2026

11281648R00069